AF576038

Nanodevices for Microwave and Millimeter Wave Applications

Nanodevices for Microwave and Millimeter Wave Applications

Special Issue Editor

Isabelle Huynen

MDPI • Basel • Beijing • Wuhan • Barcelona • Belgrade • Manchester • Tokyo • Cluj • Tianjin

Special Issue Editor
Isabelle Huynen
Université catholique de
Louvain
Belgium

Editorial Office
MDPI
St. Alban-Anlage 66
4052 Basel, Switzerland

This is a reprint of articles from the Special Issue published online in the open access journal *Micromachines* (ISSN 2072-666X) (available at: https://www.mdpi.com/journal/micromachines/special_issues/Nanodevices_Microwave_Millimeter_Wave_Applications).

For citation purposes, cite each article independently as indicated on the article page online and as indicated below:

LastName, A.A.; LastName, B.B.; LastName, C.C. Article Title. *Journal Name* **Year**, *Article Number*, Page Range.

ISBN 978-3-03936-222-6 (Hbk)
ISBN 978-3-03936-223-3 (PDF)

Contents

About the Special Issue Editor

Isabelle Huynen received the Ph.D. degree in applied sciences from the Université catholique de Louvain (UCLouvain), Louvain-la-Neuve, Belgium, in 1994. Since 1999, she has been with the National Fund for Scientific Research (FRS-FRNS), Belgium, where she is currently Research Director, and is a part-time Professor with UCLouvain. Her current research interests include nanotechnology, nanodevices, and nanomaterials for microwave and millimeter wave applications, including metamaterials, antennas, and absorbers. She (co-)authored over 130 publications in peer-reviewed journal and more than 150 communications in conference proceedings.

Editorial

Editorial for the Special Issue on "Nanodevices for Microwave and Millimeter Wave Applications"

Isabelle Huynen

Institute of Information and Communication Technologies, Electronics and Applied Mathematics (ICTEAM), Université catholique de Louvain, Place du Levant 3/P.O. Box L5.03.02, 1348 Louvain-la-Neuve, Belgium; isabelle.huynen@uclouvain.be

Received: 24 April 2020; Accepted: 26 April 2020; Published: 2 May 2020

Initially inspired by the work of Richard Feynman in 1959 during his famous talk "There is plenty of room at the bottom", nanoscience and nanotechnology have moved during the 2000s from laboratory developments to daily life applications. The nanoworld, as understood today, is at the frontier between the level of atoms and molecules, governed by quantum physics and the macroworld where materials have bulk properties resulting from the assembly of billions of atoms.

In this context, I am pleased as a Guest Editor to introduce this Special Issue focused on the interaction between nanoscale dimensions and centimeter to millimeter wavelengths. Such interaction is efficient for the design and fabrication of devices showing enhanced performances. This issue includes seven novel contributions in the field. The applications cover electronics, sensors, signal processing through amplification and the switching, mixing and broadcasting of signals, all exploiting nanoscale/nanotechnology at microwave and millimeter waves.

Heredia et al. [1] presented a frequency-reconfigurable 130nm SiGe:C BiCMOS, very compact Low Noise Amplifier (LNA) in the range 120–140 GHz. The LNA design is based on an Heterojunction Bipolar Transistor (HBT)-based switch instead of a RF-MEMS switch. A systematic procedure was applied to design the input-, inter-stage- and output-matching networks in order to obtain a perfectly balanced gain and noise figure at both frequency states. The measured gain and noise figures were 14.2/14.2 dB and 8.2/8.2 dB at 120/140 GHz, respectively, in very good agreement with circuit/electromagnetic co-simulations.

In [2] Rydosz et al. proposed a microwave gas sensor based on a 250 nm-tick SnO_2 film. Its sensitivity to acetone was improved by applying UV illumination, which emphasized the sensor's response to lower gas concentrations. As a result, detection was obtained for a wide range of gas concentrations. Various experimental conditions were tested. The highest sensitivity was obtained for a UV (375 nm) current of 10 mA at a continuous wave.

In [3] Tripon-Canseliet et al. achieved the microwave extraction of the electrical impedance of vertically aligned multi-wall carbon nanotube (VA MWCNT) bundles/forests grown on a silicon substrate. Dedicated resonating devices were designed for antenna application, operating around 10 GHz and benefiting from a natural inductive/capacitive behavior or complex conductivity in the microwave domain. As obtained from the S-parameter measurements, the capacitive and inductive behaviors of VA MWCNT bundles were deduced from the device frequency resonance shift.

Van Kerckhoven et al. [4] developed and compared two laser-assisted processes for the fabrication of microwave devices based on metallic nanowire arrays loaded inside porous alumina templates, allowing the realization of substrate integrated waveguide (SIW)-based devices. A nanowired SIW was firstly presented. It operated between 8.5 and 17 GHz, corresponding to the first and second cut-off frequencies of the waveguide, respectively. Then, a nanowired SIW isolator was demonstrated. It showed a nonreciprocal isolation of 12 dB, observed in absence of a DC magnetic field, which was achieved through an adequate positioning of ferromagnetic nanowires inside the waveguide.

In the millimeter waves range, a novel J band (i.e., between 220 and 325 GHz) MEMS switch was developed by Zhang et al. [5]. In order to improve the isolation in the "DOWN" state, the capacitance value in the "DOWN" state was increased by a thin isolation layer Si_3N_4 on the bottom plate of the switch with a thickness of ~100 nm. The switch was actuated under a voltage of ~30 V. More importantly, the switch achieved a low insertion loss of ~1.2 dB at 220 GHz and <~4 dB from 220 GHz to 270 GHz in the "UP" state, and an isolation of ~16 dB from 220 GHz to 320 GHz in the "DOWN" state.

In the field of microwave antennas Baba et al. [6] proposed a wideband planar and integrated a feeding structure for resonant cavity antennas (RCAs) including partially reflective structures, as a convenient alternative in applications requiring high directivity and bandwidth. A maximum antenna gain of 16.7 dBi was obtained with a directivity bandwidth covering nearly the entire Ku-band.

In the field of signal processing Wu et al. [7] developed a 135–190 GHz self-biased broadband frequency doubler based on planar Schottky diodes having an epitaxial layer thickness of 260 nm. The measured results showed that the doubler exhibited a 3 dB bandwidth of 34% from 135 GHz to 190 GHz, with a conversion efficiency of above 4% when supplied with 100 mW of input power. A 17.8 mW peak output power with a 10.2% efficiency was measured at 166 GHz when the input power was 174 mW, in agreement with the simulations.

I would like to take this opportunity to thank all the authors for submitting their papers to this Special Issue. I also want to thank all the reviewers for dedicating their time and helping to improve the quality of this Special Issue.

Conflicts of Interest: The author declares no conflict of interest.

References

1. Heredia, J.; Ribó, M.; Pradell, L.; Wipf, S.T.; Göritz, A.; Wietstruck, M.; Wipf, C.; Kaynak, M. Miniature Switchable Millimeter-Wave BiCMOS Low-Noise Amplifier at 120/140 GHz Using an HBT Switch. *Micromachines* **2019**, *10*, 632. [CrossRef] [PubMed]
2. Rydosz, A.; Staszek, K.; Brudnik, A.; Gruszczynski, S. Tin Dioxide Thin Film with UV-enhanced Acetone Detection in Microwave Frequency Range. *Micromachines* **2019**, *10*, 574. [CrossRef] [PubMed]
3. Tripon-Canseliet, C.; Xavier, S.; Fu, Y.; Martinaud, J.-P.; Ziaei, A.; Chazelas, J. Experimental Microwave Complex Conductivity Extraction of Vertically Aligned MWCNT Bundles for Microwave Subwavelength Antenna Design. *Micromachines* **2019**, *10*, 566. [CrossRef] [PubMed]
4. Van Kerckhoven, V.; Piraux, L.; Huynen, I. Fabrication of Microwave Devices Based on Magnetic Nanowires Using a Laser-Assisted Process. *Micromachines* **2019**, *10*, 475. [CrossRef] [PubMed]
5. Zhang, N.; Yan, Z.; Song, R.; Wang, C.; Guo, Q.; Yang, J. Design and Performance of a J Band MEMS Switch. *Micromachines* **2019**, *10*, 467. [CrossRef] [PubMed]
6. Baba, A.A.; Hashmi, R.M.; Asadnia, M.; Matekovits, L.; Esselle, K.P. A Stripline-Based Planar Wideband Feed for High-Gain Antennas with Partially Reflecting Superstructure. *Micromachines* **2019**, *10*, 308. [CrossRef] [PubMed]
7. Wu, C.; Zhang, Y.; Cui, J.; Li, Y.; Xu, Y.; Xu, R.-M. A 135-190 GHz Broadband Self-Biased Frequency Doubler using Four-Anode Schottky Diodes. *Micromachines* **2019**, *10*, 277. [CrossRef] [PubMed]

Article

Miniature Switchable Millimeter-Wave BiCMOS Low-Noise Amplifier at 120/140 GHz Using an HBT Switch

Julio Heredia [1], Miquel Ribó [2], Lluís Pradell [1,*], Selin Tolunay Wipf [3], Alexander Göritz [3], Matthias Wietstruck [3], Christian Wipf [3] and Mehmet Kaynak [3]

1 Universitat Politècnica de Catalunya, Campus Nord UPC mòdul D3, 08034 Barcelona, Catalonia, Spain
2 La Salle—Universitat Ramon Llull, 08022 Barcelona, Catalonia, Spain
3 IHP—Leibniz-Institut für innovative Mikroelektronik, 15236 Frankfurt (Oder), Germany
* Correspondence: pradell@tsc.upc.edu

Received: 24 July 2019; Accepted: 20 September 2019; Published: 21 September 2019

Abstract: A 120–140 GHz frequency-switchable, very compact low-noise amplifier (LNA) fabricated in a 0.13 μm SiGe:C BiCMOS technology is proposed. A single radio-frequency (RF) switch composed of three parallel hetero junction bipolar transistors (HBTs) in a common-collector configuration and a multimodal three-line microstrip structure in the input matching network are used to obtain a LNA chip of miniaturized size. A systematic design procedure is applied to obtain a perfectly balanced gain and noise figure in both frequency states (120 GHz and 140 GHz). The measured gain and noise figure are 14.2/14.2 dB and 8.2/8.2 dB at 120/140 GHz respectively, in very good agreement with circuit/electromagnetic co-simulations. The LNA chip and core areas are 0.197 mm^2 and 0.091 mm^2, respectively, which supposes an area reduction of 23.4% and 15.2% compared to other LNAs reported in this frequency band. The experimental results validate the design procedure and its analysis.

Keywords: low-noise amplifier (LNA); frequency-reconfigurable LNA; multimodal circuit; SiGe BiCMOS; hetero junction bipolar transistor (HBT); RF switch

---

1. Introduction

The D-band (110–170 GHz) millimeter-wave (mm-wave) frequency region is attracting a growing interest for the development of future short-distance, high-speed line-of-sight fixed-communication systems [1,2]. The International Telecommunication Union (ITU) radio regulations [3] define frequency allocations for those services in the frequency ranges 122.25–123 GHz and 141–148.5 GHz. Applications such as backhaul, front-haul and fixed wireless access in point-to-point or point-to-multipoint systems, 5G mobile backhaul tail link and inter-server connection systems have been envisaged, and several demonstration prototypes of D-band links in the frequency range 125–160 GHz have been reported [4]. A multiband architecture is often required for an efficient use of the spectrum [5,6] and, in order to enable inexpensive and versatile communication systems, frequency reconfigurable devices, in particular low-noise amplifiers (LNAs) are highly desirable. This way the size, cost, and power consumption are minimized.

The SiGe BiCMOS technology has demonstrated maturity and high performance for D-band wireless systems and sensors [2,7,8]. Recent advances in 0.13 μm SiGe BiCMOS hetero junction bipolar transistor (HBT) devices [9] have enabled the development of fixed-frequency LNAs using two to four HBT stages, typically in a cascode configuration [1,10–17], featuring a high gain G (G = 24 dB at 158 GHz [1]) and low noise figure F (F = 5.1 dB at 144.5 GHz [14]). Other silicon technologies, such as 65- and 28-nm SOI CMOS, are also being used to fabricate D-band LNAs [18–20], featuring comparable gain but a higher noise figure (G = 15.7 dB and F = 8.5 dB at 160 GHz [20]). Regarding frequency-reconfigurable mm-wave LNAs, only a few have been reported to date, at 60/77 GHz [15], 24/79 GHz [16], and at D-band (125/140 GHz) [17], all using the SiGe BiCMOS process. The frequency reconfiguration is provided with radio-frequency microelectromechanical system (RF-MEMS) switches (elements included in the design-kit library [8]). Though the design in [17] was not optimized for a maximal G or minimal F, but for a balanced G and F in the two frequency states, it features a gain G = 18.2/16.1 dB and noise figure F = 7/7.7 dB at 125/140 GHz, which compare well with those of fixed-frequency LNAs [1,11,12] and are competitive for the intended applications.

Regarding miniaturization, the frequency-reconfigurable D-band LNA presented in [17] exhibits both, the smallest chip area, A_{CHIP}, and core area (without RF- and DC-pads), A_{CORE}, compared to the other D-band LNAs [1,10–14]. In [17], the RF-MEMS switch area is 0.031 mm^2, which supposes a 29% of A_{CORE} (A_{CORE} = 0.107 mm^2). Since the RF-MEMS switch dimensions barely scale with frequency (0.040 mm^2 for devices in the 60–110 GHz [15] vs. 0.031 mm^2 for devices in the 110–170 GHz frequency band [17]), it is pertinent to explore other options to design more compact RF switches. One potential application is future frequency-reconfigurable BiCMOS LNAs at higher frequencies, such as the G-band (140–220 GHz) and WR-4 waveguide band (170–260 GHz), at which the RF-MEMS switch area may become comparable to the core area. As an example, the (estimated) A_{CORE} of the 245 GHz fixed-frequency LNA reported in [21] is only 0.036 mm^2.

In this paper a frequency-reconfigurable BiCMOS LNA at 120/140 GHz is presented. The LNA design is based on the one reported by the authors in [17] but now an HBT-based switch is used instead of a RF-MEMS switch. As a result, the LNA size is further reduced with only a slight decrease in gain and increase in noise figure, but with enhanced gain balance and noise-figure balance in both frequency states. The paper is organized as follows: In Section 2 the LNA design and HBT-based switch design are presented, and the HBT-switch performances are compared to that of the RF-MEMS switch. In Section 3 the experimental results are presented, discussed, and compared to previous works. Finally, Section 4 concludes the paper and its contributions are highlighted.

2. Design of the Frequency-Reconfigurable LNA

The LNA schematic is shown in Figure 1. It consists of two cascode stages. The first stage is composed of HBTs Q_1 and Q_2 and the second stage of HBTs Q_3 and Q_4. They are biased using current mirrors (Q_5/Q_6 and Q_7/Q_8, respectively) and bias resistors (R_1 and R_2). Similarly to [17], the input and output matching networks (IMN and OMN, respectively) are fixed, and the frequency-reconfiguration capability is only introduced in the LNA inter-stage matching network (ISMN), to reduce the design complexity, size, and weight.

The IMN is based on a compact three-line-microstrip (TLM) section connected to input/output short microstrip lines, with two central series gaps in its outer strips [17]. The TLM simultaneously propagates three fundamental modes: even-even (*ee*), odd-odd (*oo*), and odd-even (*oe*) [22]. The microstrip modes basically generate *ee* modes at the microstrip-to-TLM transitions, which then excite (and afterwards interact with) the *oo* and *oe* modes at the gaps, which in turn resonate in the TLM section. The interaction between all these modes results in an IMN featuring a large electrical size with reduced physical area. To illustrate the modal conversion as well as the resonance condition in the TLM multimodal structure, Figure 2 shows the simulated current density distribution on the metallization for the two frequency states (120 GHz and 140 GHz). The simulation was performed using the electromagnetic (EM) simulator Momentum 2.5D (Keysight Technologies, Santa Rosa, CA, USA). The (symmetrical) *ee* mode features equally-oriented currents on the three TLM strips, the (symmetrical) *oo* mode features equally-oriented currents in the outer strips and reverse-oriented current in the center strip, and the (anti-symmetrical) *oe* mode features reverse-oriented currents in the outer strips and no current in the center strip [19]. The anti-symmetrical *oe* mode can only be generated in the asymmetrically loaded gaps in the outer TLM strips, whereas the symmetrical *ee* and *oo* modes can, in addition, be generated at the symmetrical transitions to microstrip lines at both ports of the IMN. The high levels of currents in the center strip for the 120-GHz state indicate the predominance of the *ee* and *oo* modes for the performance of the IMN in this state, whereas the markedly asymmetrical current distribution for the 140-GHz state indicates the predominance of *oo* and *oe* modes in the performance of the IMN in this state. The characteristic impedances and propagation constants of each TLM-fundamental mode were first determined from electromagnetic simulation of a basic TLM section without any discontinuity. Then the dimensions (lengths, strip widths, and gaps) of the TLM structure in the IMN were optimized by using modal equivalent circuits which transform the actual voltages and currents on generic n-conductor lines (in this case, n = 3) into their modal counterparts [23]. This way the IMN physical dimensions were minimized to obtain a structure of reduced size designed to combine a low LNA noise figure (F) and a small input-reflection coefficient magnitude $|\Gamma_{IN}|$, both balanced at the two frequency states (120/140 GHz).

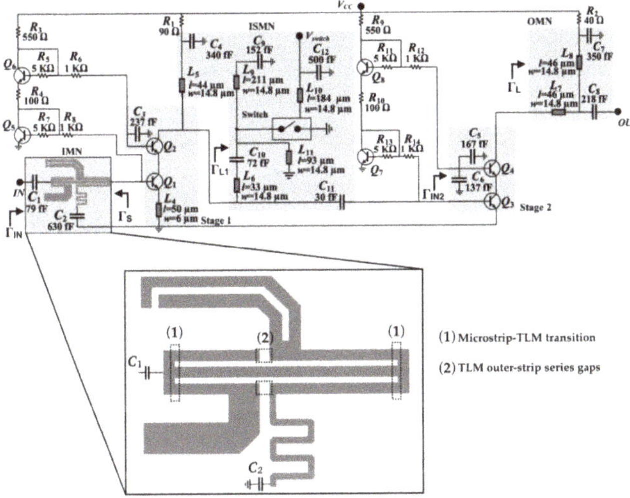

Figure 1. Schematic of the switchable BiCMOS low-noise amplifier (LNA) at 120/140 GHz and a detail of the multimodal three-line-microstrip (TLM) structure implemented in the input matching networks (IMN).

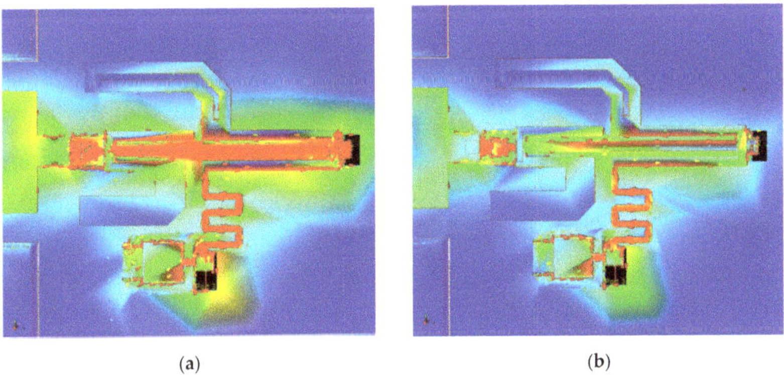

(**a**) (**b**)

Figure 2. IMN electromagnetic (EM) simulation results. Simulated current distribution on the IMN TLM multimodal structure. (**a**) 120 GHz. (**b**) 140 GHz.

The OMN is composed of a simple line-short-circuited stub structure (microstrip lines L_7 and L_8), designed to attain a balanced second-stage power gain G_{p2} at both frequency states. The ISMN consists of a two-segment short-circuited stub L_6/L_9, a second short-circuited stub L_5, an RF switch terminated with a short-circuited stub L_{11}, biased through a λ/4 stub (L_{10}), and a series capacitor C_{11}. C_9, C_{10}, and C_{12} are decoupling capacitors. The purpose of L_5 is to allow a shorter L_6 segment, thus reducing the chip transversal dimension. The RF switch selects the two-segment stub length, either L_6 when the switch is in its ON state (for the high-frequency, 140 GHz LNA state), or $L_6 + L_9$ when the switch is in its OFF state (for the low-frequency, 120 GHz LNA state). As shown in Figure 3, the ISMN line lengths and the series capacitor value C_{11} were designed to present Γ_{L1} values that guarantee a first-stage power gain G_{p1} balanced at both frequency states (G_{p1} = 9.6/8.9 dB at 120/140 GHz). Therefore, the combined effect of the OMN and the ISMN is to balance the LNA power gain G_p ($G_p = G_{p1} \cdot G_{p2}$) at both frequency states. Those Γ_{L1} values simultaneously fulfill the IMN requirements for a low and balanced F and $|\Gamma_{IN}|$. This can graphically be assessed by plotting on the Γ_{L1} plane the geometrical locus of constant F and $|\Gamma_{IN}|$ (for a given source reflection coefficient), which is a circle. These circles, for $|\Gamma_{IN}| = -11/-14.6$ dB and F = 6.1/6.5 dB at 120/140 GHz, are also plotted in Figure 3. It can be observed that the circles intersect the intended Γ_{L1} values, and in consequence the requirements for G_p, F, and $|\Gamma_{IN}|$ at both frequency states are satisfied simultaneously. The simulated results were obtained from circuit/electromagnetic (EM) co-simulation, using manufacturer circuit models for HBTs (including the transistors used in the RF switch) and resistors, and electromagnetic models (obtained from electromagnetic simulation) for passives (lines, metallization on different layers, pads, via-holes, capacitors and ground plane). The simulation platform was Keysight ADS/2.5D Momentum. From the electromagnetic analysis an electromagnetic model (co-simulation element) was obtained, which was used in a circuit simulation as a block to which the manufacturer's design kit models of HBTs and resistors were connected.

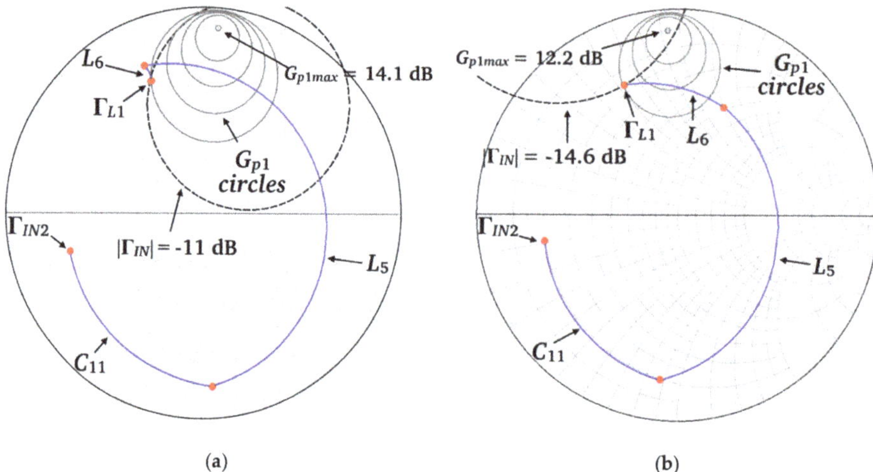

Figure 3. Inter-stage matching network (ISMN) design and plots of constant G_{p1} circles and constant F and $|\Gamma_{IN}|$ circles on the Γ_{L1} plane. (**a**) 120 GHz frequency state; (**b**) 140 GHz frequency state.

For the RF switch, three different configurations were considered (shown in Figure 4a–c). The shunt-HBT configuration (Figure 4a) uses three shunt-connected HBTs in reverse-saturation mode. The diode configuration (Figure 4b) uses one or two shunt-connected HBTs as diodes. The L-shape configuration (Figure 4c) uses two HBTs, series- and shunt-connected respectively. Preliminary simulations showed that the L-shape and shunt-HBT configurations had similar ON-state isolation but the L-shape OFF-state insertion loss was much higher. The diode configuration showed poorer ON-state isolation and OFF-state insertion loss than the shunt-HBT configuration. Therefore, the L-shape and diode configurations were dismissed as an option for the RF switch, and the selected configuration was the shunt-HBT in reverse-saturation mode (Figure 4a). The three shunt HBTs (Q_1, Q_2, and Q_3 in Figure 4a) are in a common-collector configuration. With reference to Figure 1, the emitters are connected to L_{11}, the collectors are connected to ground and the switch bias voltage V_{SWITCH} is supplied through the λ/4 stub L_{10}. When V_{SWITCH} = 0 V (Figure 4b), the transistors are in the cut-off region mode and they behave as a high impedance R_{OFF} in parallel with a parasitic capacitance C_{OFF} (the HBT-based RF switch is in OFF state). L_{11} adds an inductance which resonates with the C_{OFF} capacitances, thus reducing their effect on the switch. When V_{SWITCH} = 1 V (Figure 4c), the transistors are in the saturation region mode and they behave as a low impedance R_{ON} (the HBT-based RF switch is in ON state). Connecting the switch HBTs in reverse saturation mode (the HBTs are flipped) isolates the emitter from the silicon substrate, thus reducing the parasitic capacitances; also, the impedance in OFF state is larger since the potential barrier in the conduction band is larger at the emitter than at the collector [24]. The RF-switch DC-power consumption in ON state is 15 mW. The number of transistors used for the selected shunt-HBT configuration was a trade-off between performance and power consumption. If two HBTs were used, the DC-power consumption would decrease to 12 mW but the ON-state isolation would worsen (15.5 dB). Using four HBTs would increase the ON-state isolation to 20.8 dB but the OFF-state insertion loss and power consumption would increase to 0.46 dB and 24 mW, respectively.

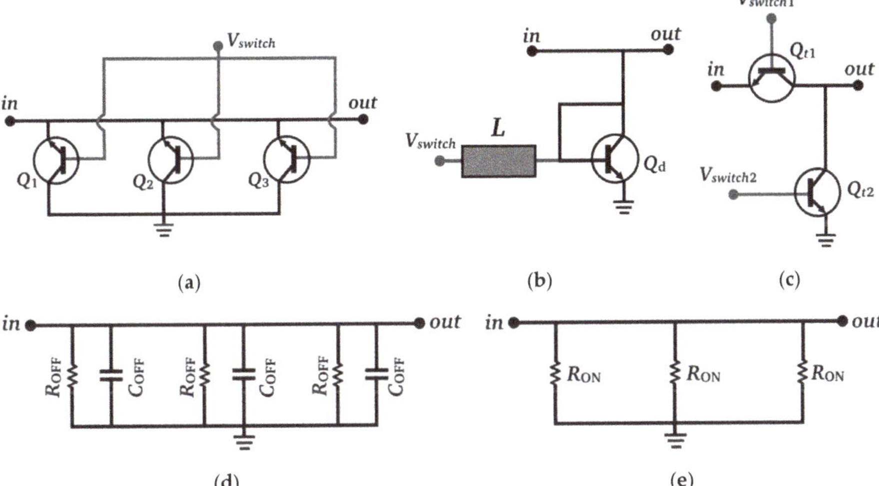

Figure 4. Hetero junction bipolar transistor (HBT) radio frequency (RF) switch. (**a**) Selected witch configuration (shunt-connected HBTs in reverse saturation mode). The switch activation voltage is V_{SWITCH} = 1 V. Other switch configurations: (**b**) Diode configuration; (**c**) L-shape configuration. Equivalent circuit of the selected switch (**a**): (**d**) OFF state (C_{OFF} = 7.2 fF, R_{OFF} = 1600 Ω); (**e**) ON state (R_{ON} = 10 Ω).

Figure 5 plots simulations of the proposed HBT-switch *S*-parameters for the configuration shown in Figure 4a, taking into consideration the stubs L_{10} and L_{11} connected to the HBT base and emitter, respectively, as shown in Figure 1. From Figure 5, the calculated switch-insertion loss (*IL*), defined as $IL = -10\cdot\log(|S_{21}(\mathrm{OFF})|^2)$, is 0.36 dB at 120 GHz, and the calculated switch isolation (*I*), defined as $I = -10\cdot\log(|S_{21}(\mathrm{ON})|^2)$, is 18.6 dB at 140 GHz. The stub length L_{11} was tuned to resonate with C_{OFF} to minimize *IL* at 120 GHz. The RF-MEMS switch used in [17] features *IL* = 0.25 dB at 120 GHz and *I* = 32 dB at 140 GHz [8]. Therefore, the proposed HBT switch presents a lower isolation and a higher insertion loss than its RF-MEMS switch counterpart. The performance trade-off between MEMS and HBT switches for the LNA design can be observed by comparing Figure 3 with Figure 2 of [17]. G_{p1} decreases and the noise figure increases in both frequency states, but the decrease in G_{p1} is smaller in the high-frequency state (1.6/1 dB respectively). Thus, as discussed in Section 3, it is possible to get a perfectly balanced gain and noise figure between both frequency states, and the chip size is substantially reduced.

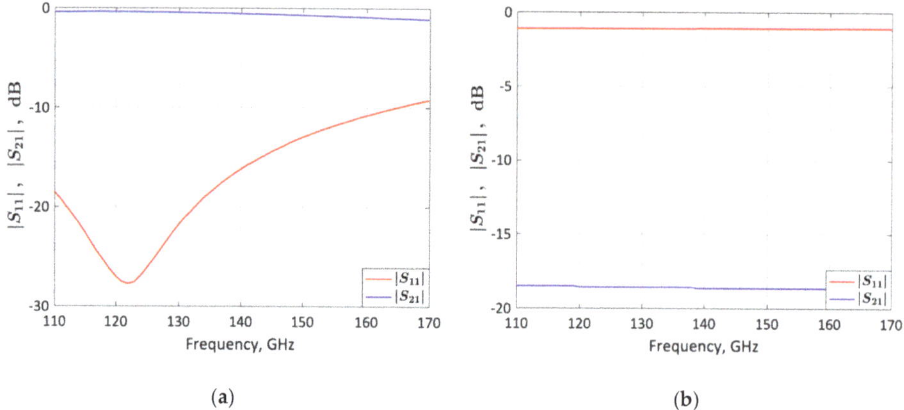

Figure 5. Simulated *S*-parameters of the HBT-RF switch of Figure 4a (designed on SG13G2 0.13 μm SiGe:C BiCMOS technology) taking into consideration the stubs L_{10} and L_{11} connected to the HBT base and emitter, respectively, as shown in Figure 1. (**a**) Switch in OFF state. (**b**) Switch in ON state.

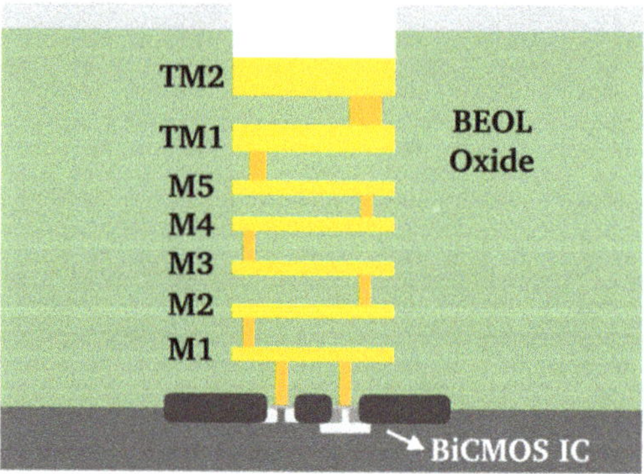

Figure 6. Cross section of the back-end-of-line (BEOL) in the SG13G2 0.13 μm SiGe:C BiCMOS technology from IHP—Leibniz-Institut für innovative Mikroelektronik [25].

The LNA was fabricated in SG13G2 0.13 μm SiGe:C BiCMOS technology using HBTs with f_T/f_{max} of 300/500 GHz and 0.9 μm emitter length [25] from IHP—Leibniz-Institut für innovative Mikroelektronik. The back-end-of-line (BEOL) consists of five metal layers (M1–M5) and two top-metal layers, TM1, and TM2 (Figure 6). All lines on the ISMN (L_5, L_6, L_9, L_{10}, and L_{11}) and OMN (L_7 and L_8,), as well as line L_4 are microstrip lines. The top-most metal layer TM2 was used for L_5, L_6, L_7, L_8 L_{10}, L_{11}, and the multimodal TLM structure of the IMN. L_9 was fabricated using the TM1 layer, and L_4 with a stack of three layers (M2, M3, and M4). The calculated Q factor from EM simulation is 37 for both frequencies (120 GHz and 140 GHz). The number of emitter fingers for transistors $Q_1/Q_2/Q_3/Q_4$ in stages 1 and 2 of the cascode configuration (Figure 1) is 5/10/10/10. This combination was selected according to the gain and noise figure required for each stage, while keeping a low DC-power consumption. The number of emitter fingers of each transistor in the HBT-RF switch is 7, which assures the required current flows for the transistors in ON state (I_B = 5 mA for each transistor, I_C = 1.9/1.7/1.7 mA for $Q_1/Q_2/Q_3$ in

Figure 4a). The LNA total DC-power consumption P_{DC}, including the two stages and the RF switch is P_{DC} = 37.5 mW for the 120 GHz frequency state and P_{DC} = 52.5 mW for the 140 GHz frequency state.

Figure 7 shows a micrograph of the fabricated LNA. The LNA features an HBT-switch area of 45.4 μm × 23.6 μm, which is much smaller than the RF-MEMS switch area (260 μm × 118 μm) of the previous LNA presented in [17]. The chip and core areas are A_{CHIP} = 515 μm × 382 μm and A_{CORE} = 331 μm × 274 μm, which supposes an area reduction of 23.4% and 15.2%, respectively, compared to [17].

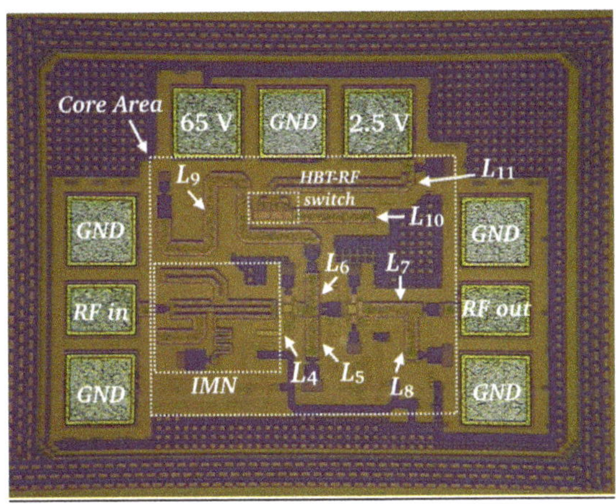

Figure 7. Micrograph of the frequency-switchable LNA including and HBT-RF switch fabricated in a SG13G2 0.13 μm SiGe:C BiCMOS technology from IHP [25]. Dimensions are: A_{CHIP} = 515 μm × 382 μm and A_{CORE} = 331 μm × 274 μm. The HBT-RF switch area is: 45.4 μm × 23.6 μm.

3. Results and Discussion

3.1. S-Parameter Simulation and Measurement

The S-parameters of the fabricated frequency-switchable LNA were experimentally characterized from 110 to 170 GHz on a semi-automated wafer probe station with a setup from Rohde & Schwarz, consisting of a 4 port ZVA24 as vector network analyzer and two ZVA170 millimeter-wave converters (Figure 8). The Cascade Microtech 75 μm-pitch infinity(R) GSG waveguide probes were connected via WR6 waveguide s-bends to the millimeter-wave converters. For the calibration, the impedance standard substrate (ISS 138–356) was placed together with an RF absorber on an auxiliary ceramic chuck and a full two-port LRRM calibration was performed. The applied bias voltage V_{CC} (Figure 1) was V_{CC} = 2.5 V and the input RF power was −20 dBm. Figures 9 and 10 compare the measured and simulated LNA S-parameters for the lower-frequency state (120 GHz) and upper-frequency state (140 GHz), respectively. The LNA features a measured $|S_{21}|$, $|S_{11}|$, $|S_{12}|$, and $|S_{22}|$ of 14.2, −6.6, −46, and −8.1 dB respectively for the lower-frequency state, and 14.2, −14, −37.9, and −2.5 dB respectively for the upper-frequency state. These results are in good agreement with simulations.

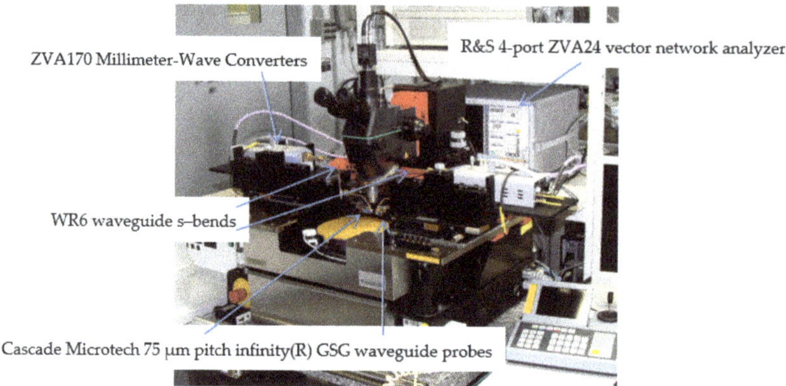

Figure 8. 110–170 GHz S-parameter measurement setup using a semi-automated wafer probe station.

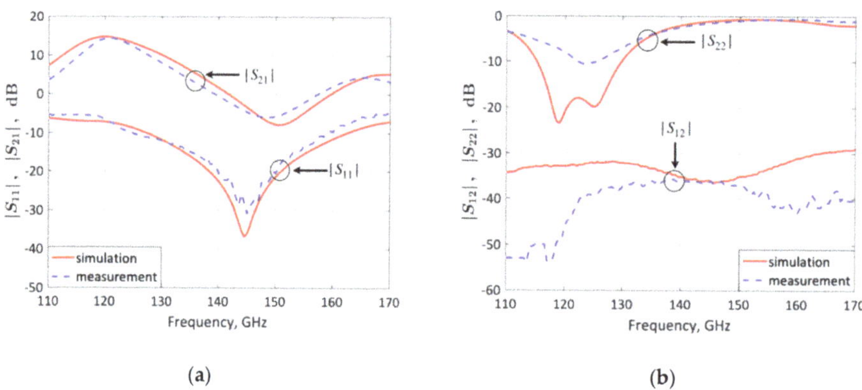

Figure 9. Measured and simulated LNA S-parameters for the lower-frequency state (120 GHz). (**a**) S_{21} and S_{11}; (**b**) S_{12} and S_{22}.

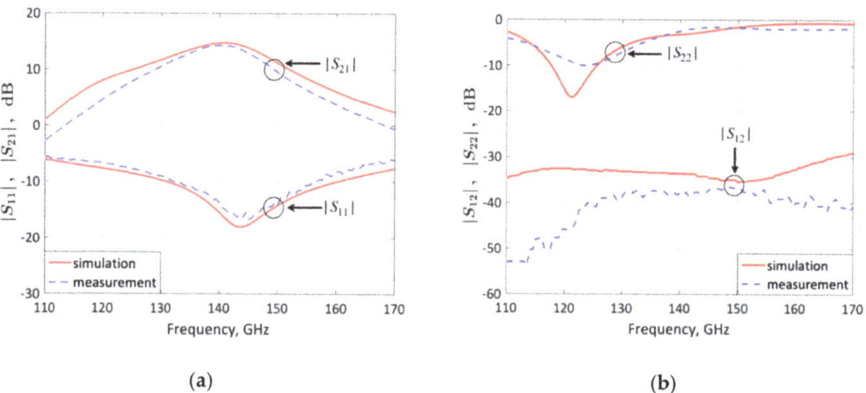

Figure 10. Measured and simulated LNA S-parameters for the upper-frequency state (140 GHz). (**a**) S_{21} and S_{11}; (**b**) S_{12} and S_{22}.

3.2. Noise-Figure (F) Simulation and Measurement

The noise figure F was measured for both frequency states. The measurement was also carried out on wafer using the Y-factor method. The measurement setup is described in [20]. Hot and cold noise temperatures are produced by a noise diode Elva-1 ISSN-06. The noise power is down-converted to a 50 MHz IF using a subharmonic mixer with amplifier-multiplier chain as LO (MixAMC-192, Virginia Diodes Inc., Charlottesville, VA, USA), and measured using a noise figure analyzer Agilent N8973A (now Keysight Santa Rosa, CA, USA).

The simulated F is compared to the measured results for the lower- and upper-frequency states in Figure 11. The measured F is 8.2 dB for both lower- and upper-frequency states, in close agreement with simulation.

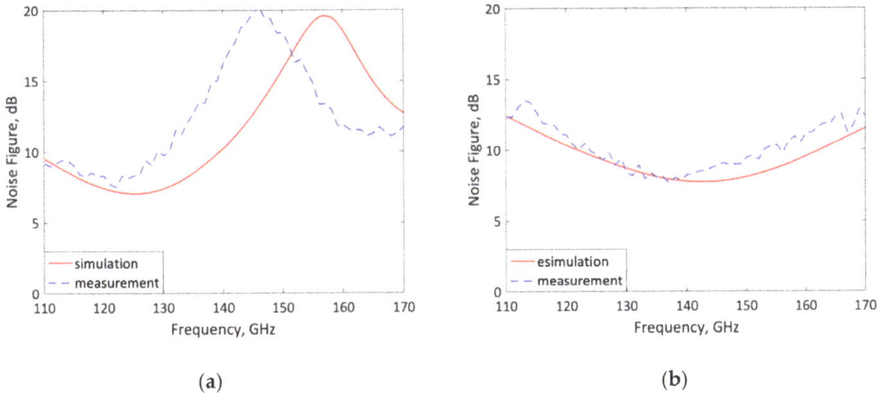

(a) (b)

Figure 11. Measured and simulated LNA noise figure. (**a**) Lower-frequency state (120 GHz); (**b**) upper-frequency state (140 GHz).

The noise-figure peak shift in low-frequency state is attributed to the EM simulation which underestimates the small inductance associated to the metallization and via holes connecting each transistor emitter of the HBT switch.

3.3. Stability (μ-Factor) Simulation and Measurement

The stability of the LNA was assessed using the μ and μ' factors [26]. According to the simulations performed from DC to 170 GHz, the LNA is unconditionally stable in all frequencies ($\mu > 1$ and $\mu' > 1$) for both lower- and upper-frequency states. The μ and μ' stability factors were also obtained from the measured results demonstrating that the LNA is unconditionally stable in the 110 to 170 GHz frequency band for both frequency states. As shown in Figure 12, the minimal calculated μ and μ' values obtained from the measurements are $\mu = 1.07$ and $\mu' = 1.82$ for the lower- frequency state, and $\mu = 1.02$ and $\mu' = 1.45$ for the upper-frequency state.

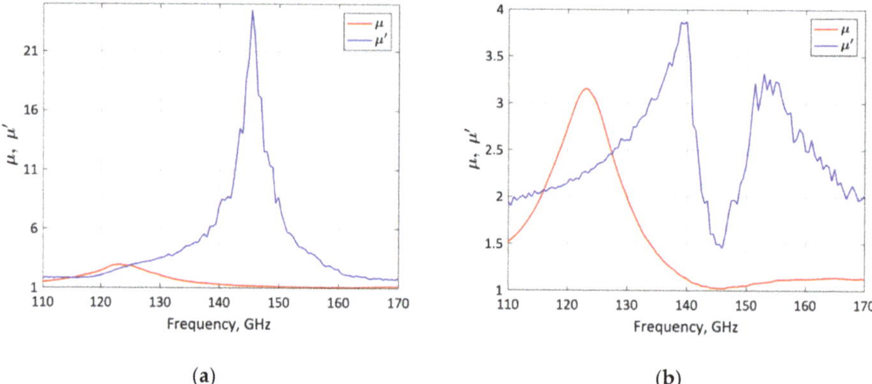

Figure 12. Stability factors (μ and μ') obtained from the measured S-parameters. (**a**) Lower-frequency state (120 GHz); (**b**) upper-frequency state (140 GHz).

3.4. Simulated 1-dB Gain Compression Point (P_{1dB})

The 1-dB gain compression point (P_{1dB}) was obtained from a non-linear simulation of the LNA. The LNA input power was swept from −50 dBm to 0 dBm and the output power was simulated using the HBT non-linear model provided by the manufacturer. In Figure 13, the simulated output power is plotted vs. the input power and compared to the theoretical small-signal output power. It can be observed that the input-referred P_{1dB} is −12.4 dBm for the lower-frequency state and−13.6 dBm for the upper-frequency state.

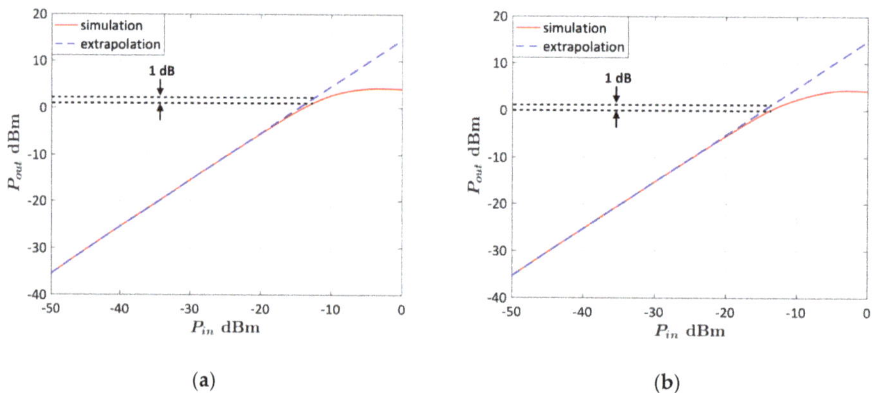

Figure 13. Non-linear simulation of the LNA output power vs. input power. (**a**) Lower-frequency state (120 GHz); (**b**) upper-frequency state (140 GHz).

3.5. Discussion of Results

The LNA S-parameters and F measurements presented in Figures 9–11 are in good agreement with the simulations and well balanced at both frequency states. Indeed, it is observed that the gain ($|S_{21}|$) and noise figure F feature the same measured values (14.2/14.2 dB and 8.2/8.2 dB) at 120/140 GHz, respectively. Thus, the frequency-reconfigurable LNA concept and design methodology is validated. Table 1 shows a comparison of the LNA measured parameters and those of other reconfigurable and not-reconfigurable cascaded SiGe BiCMOS mm-wave LNAs reported in the literature. In order to evaluate and compare the LNA performance, a figure-of-merit (FoM) is defined as

$$FoM = \frac{1000 \times G \cdot P_{1dB}}{(F-1) \cdot P_{DC} \cdot A} \tag{1}$$

where G is the LNA gain ($G = |S_{21}|^2$) and the area A refers either to A_{CHIP} or A_{CORE}.

The proposed LNA exhibits the smallest area A (both A_{CHIP} and A_{CORE}) and the highest FoM (save [11] and [17] at 143 GHz with a similar FoM). Compared to the frequency-reconfigurable LNAs based on RF-MEMS switches [15–17], it is more compact at the expense of a lower gain and higher P_{DC} in the upper-frequency state, and it exhibits the same (or very similar) F at comparable frequencies. Though the LNA was not designed for a maximal G or minimal F, but for a balanced G and F in the two frequency states, its F is comparable or better than [1,12] at similar frequencies, which are not reconfigurable and were optimized for low-noise performance.

Table 1. Comparison with other cascaded SiGe BICMOS mm-wave LNAs.

	Technology (μm)	Frequency (GHz)	G (dB)	P_{1dB} (dBm)	F (dB)	P_{DC} (mW)	A_{CHIP} / A_{CORE} (mm^2)	FoM
[1] †	0.13	158	24.1	−25.9	8.2	28	0.342/0.18††	12.31/23.38
[10]	0.13	140	23.3	−33**	5.5	12	0.393/0.231††	8.92/15.17
[11]	0.13	130	24.3	−17.3	6.8	84	0.301/0.192††	52.35/82.07
[12]	0.13	145	21		8.5	14.5	0.36/0.270††	
[13]	0.09	140	30		6.2	45	0.525/0.115	
[14]	0.13	144.5	32.6	−37.6	5.1++	28	1/0.6	5.05/8.42
[15]	0.25	60	20*	−18**	7*	40	0.788/0.317††	12.53/31.2
[15]	0.25	77	22*	−18**	8*	40	0.788/0.317††	15.01/37.31
[16]	0.25	24	25	−27**	4.3**	40	0.770/0.476††	12.11/19.6
[16]	0.25	74	18	−18**	8.5**	40	0.770/0.476††	5.34/8.63
[17]	0.13	125	18.2	−17.3**	7	36.8	0.257/0.107	32.42/78.17
[17]	0.13	143	16.1	−15.9**	7.7	36.8	0.257/0.107	22.65/54.6
This	**0.13**	**120**	**14.2**	**−12.4****	**8.2**	**37.5**	**0.197/0.091**	**37.34/80.99**
This	**0.13**	**140**	**14.2**	**−13.6****	**8.2**	**52.5**	**0.197/0.091**	**20.12/43.63**

† Cascode; * Averaged; ** Simulated; ++ Estimated.

4. Conclusions

A 120–140 GHz frequency-reconfigurable 0.13 μm SiGe:C BiCMOS, very-compact D-band LNA has been presented. A single HBT switch is used in the inter-stage matching network to minimize size and design complexity. The HBT switch is composed of three transistors in parallel, featuring 0.36 dB switch-insertion loss in OFF state and 18.6 dB switch-isolation in ON state. The LNA size is minimized by using, in addition to a single HBT switch, a multimodal three-line-microstrip input-matching network. A systematic general procedure has been applied to design the input-, inter-stage- and output-matching networks to obtain a perfectly balanced gain and noise figure at both frequency states. The measured gain and noise figure are 14.2/14.2 dB and 8.2/8.2 dB at 120/140 GHz, respectively, in very good agreement with circuit/electromagnetic co-simulations. The chip and core areas (0.197/0.091 mm^2) are the smallest reported in the literature in this frequency band. The experimental results validate the design procedure and its analysis, and prove that reconfigurable devices based on HBTs can be a viable alternative to those based on MEMS switches whenever the performance specifications are not exceedingly demanding and compactness can be an issue (e.g., in space applications), besides saving cost and fabrication steps.

Author Contributions: Conceptualization, J.H., M.R., L.P. and M.K.; methodology, J.H., M.R. and L.P.; validation, S.T.W., A.G., M.W. and C.W.; formal analysis, J.H., M.R. and L.P.; investigation, S.T.W., A.G., M.W. and C.W.; resources, S.T.W., A.G., M.W., C.W. and M.K.; writing—original draft preparation, J.H., M.R. and L.P.; writing—review and editing, L.P., M.R., S.T.W. and M.K.; supervision, L.P., M.R. and M.K.

Funding: This research was funded by the Spanish MEC under Projects TEC2013-48102-C2-1/2-P and TEC2016-78028-C3-1-P, by the Spanish AEI under Grant Unidad de Excelencia Maria de Maeztu MDM-2016-0600, by the Catalan AGAUR under Grant 2017-SGR-00219, and by the Mexican CONACYT under Fellowship 410742.

Acknowledgments: The authors would like to thank Mikko Kantanen, MilliLab-VTT, for the noise-figure measurement and setup details.

Conflicts of Interest: The authors declare no conflict of interest.

References

1. Coen, C.T.; Ulusoy, A.Ç.; Song, P.; Ildefonso, A.; Kaynak, M.; Tillack, B.; Cressler, J.D. Design and On-Wafer Characterization of G-Band SiGe HBT Low-Noise Amplifiers. *IEEE Trans. Microw. Theory Techn.* **2016**, *64*, 3631–3642. [CrossRef]
2. Valenta, V.; Schumacher, H.; Tolunay Wipf, S.; Wietstruck, M.; Göritz, A.; Kaynak, M.; Winkler, W. Single-chip Transmit-Receive Module with a Fully Integrated Differential RF-MEMS Antenna Switch and a High-Voltage Generator for F-Band Radars. In Proceedings of the 2015 IEEE Bipolar/BiCMOS Circuits and Technology Meeting (BCTM), Boston, MA, USA, 26–28 October 2015; pp. 40–43.
3. ITU Radio Regulations. Available online: https://www.itu.int/pub/R-REG-RR (accessed on 17 July 2019).
4. Millimetre Wave Transmission (mWT); Analysis of Spectrum, License Schemes and Network Scenarios in the D-band. Available online: http://www.etsi.org/standards-search (accessed on 17 July 2019).
5. Fujimoto, R.; Motoyoshi, M.; Takano, K.; Fujishima, M. A 120 GHz/140 GHz Dual-Channel OOK Receiver Using 65 nm CMOS Technology. *IEICE Trans. Fundam. Electron. Commun. Comput. Sci.* **2013**, *E96-A*, 486–493. [CrossRef]
6. Leufker, J.D.; Fritsche, D.; Tretter, G.; Carta, C.; Ellinger, F. Dualband 180 GHz and 205 GHz Medium-Power High-Gain Amplifier on 130 nm BiCMOS. In Proceedings of the 2016 21st International Conference on Microwave, Radar and Wireless Communications (MIKON), Krakow, Poland, 9–11 May 2016.
7. Kueppers, S.; Cetinkaya, H.; Pohl, N. A Compact 120 GHz SiGe:C based 2 × 8 FMCW MIMO Radar Sensor for Robot Navigation in Low Visibility Environments. In Proceedings of the 14th European Radar Conference (EuRAD), Nuremberg, Germany, 11–13 October 2017; pp. 122–125.
8. Tolunay Wipf, S.; Göritz, A.; Wietstruck, M.; Wipf, C.; Tillack, B.; Kaynak, M. D–Band RF–MEMS SPDT Switch in a 0.13 μm SiGe BiCMOS Technology. *IEEE Microw. Wireless Compon. Lett.* **2016**, *26*, 1002–1004. [CrossRef]
9. Heinemann, B.; Rücker, H.; Barth, R.; Bärwolf, F.; Drews, J.; Fischer, G.G.; Fox, A.; Fursenko, O.; Grabolla, T.; Herzel, F.; Katzer, J.; et al. SiGe HBT with f_T/f_{max} of 505 GHz/720 GHz. In Proceedings of the 2016 International Elctron Device Meeting (IEDM), San Francisco, CA, USA, 3–7 December 2016; pp. 3.1.1–3.1.4.
10. Ulusoy, A.Ç.; Song, P.; Khan, W.T.; Kaynak, M.; Tillack, B.; Papapolymerou, J.; Cressler, J.D. A SiGe D-Band Low-Noise Amplifier Utilizing Gain-Boosting Technique. *IEEE Microw. Wireless Compon. Lett.* **2015**, *25*, 61–63. [CrossRef]
11. Hou, D.; Xiong, Y.-Z.; Goh, W.-L.; Hong, W.I.; Madihian, M. A D-Band Cascode Amplifier With 24.3 dB Gain and 7.7 dBm Output Power in 0.13 μm SiGe BiCMOS Technology. *IEEE Microw. Wireless Compon. Lett.* **2012**, *22*, 191–193. [CrossRef]
12. Zhang, B.; Xiong, Y.-Z.; Wang, L.; Hu, S.; Li, L.-W. Gain-enhanced 132–160 GHz low-noise amplifier using 0.13 μm SiGe BiCMOS. *Electron. Lett.* **2012**, *48*, 257–259. [CrossRef]
13. Yishay, R.B.; Shumaker, E.; Elad, D. A 122-150 GHz LNA with 30 dB gain and 6.2 dB noise figure in SiGe BiCMOS technology. In Proceedings of the IEEE 15th Topical Meeting on Silicon Monolithic Integrated Circuits in RF Systems (SiRF), San Diego, CA, USA, 26–28 January 2015; pp. 15–17.
14. Turkmen, E.; Burak, A.; Guner, A.; Kalyoncu, I.; Kaynak, M.; Gurbuz, Y. A SiGe HBT D-Band LNA With Butterworth Response and Noise Reduction Technique. *IEEE Microw. Wireless Compon. Lett.* **2018**, *28*, 524–526. [CrossRef]
15. Ulusoy, A.Ç.; Kaynak, M.; Purtova, T.; Tillack, B.; Schumacher, H. A 60 to 77 GHz Switchable LNA in an RF-MEMS Embedded BiCMOS Technology. *IEEE Microw. Wireless Compon. Lett.* **2012**, *22*, 430–432. [CrossRef]
16. Ulusoy, A.Ç.; Kaynak, M.; Purtova, T.; Tillack, B.; Schumacher, H. 24 to 79 GHz frequency band reconfigurable LNA. *Electron. Lett.* **2012**, *48*, 1598–1600. [CrossRef]

17. Heredia, J.; Ribó, M.; Pradell, L.; Tolunay Wipf, S.; Göritz, A.; Wietstruck, M.; Wipf, C.; Kaynak, M. A 125–143-GHz Frequency-Reconfigurable BiCMOS Compact LNA Using a Single RF-MEMS Switch. *IEEE Microw. Wireless Compon. Lett.* **2019**, *29*, 339–341. [CrossRef]
18. Dafna, Y.; Cohen, E.; Socher, E. A wideband 95–140 GHz high efficiency PA in 28 nm CMOS. In Proceedings of the IEEE 28-th Convention of Electrical and Electronics Engineers in Israel (IEEEI), Eliat, Israel, 3–5 December 2014.
19. Kim, D.-H.; Kim, D.; Rieh, J.S. A D-Band CMOS Amplifier With a New Dual-Frequency Interstage Matching Technique. *IEEE Trans. Microw. Theory Tech.* **2017**, *65*, 1580–1588. [CrossRef]
20. Parveg, D.; Varonen, M.; Karaca, D.; Vahdati, A.; Kantanen, M.; Halonen, K.A.I. Design of a D-Band CMOS Amplifier Utilizing Coupled Slow-Wave Coplanar Waveguides. *IEEE Trans. Microw. Theory Tech.* **2018**, *66*, 1359–1373. [CrossRef]
21. Schmalz, K.; Borngräber, J.; Mao, Y.; Rücker, H.; Weber, R. A 245 GHz LNA in SiGe Technology. *IEEE Microw. Wireless Compon. Lett.* **2012**, *22*, 533–535. [CrossRef]
22. Tripathi, K.V. Asymmetric Coupled Transmission Lines in an Inhomogeneous Medium. *IEEE Trans. Microw. Theory Tech.* **1975**, *23*, 734–739. [CrossRef]
23. Scanlan, J.O. Theory of Microwave Coupled-Line Networks. *Proc. IEEE* **1980**, *68*, 209–231. [CrossRef]
24. Schmid, R.L.; Ulusoy, A.Ç.; Song, P.; Cressler, J.D. A 94 GHz, 1.4 dB Insertion Loss Single-Pole Double-Throw Switch Using Reverse-Saturated SiGe HBTs. *IEEE Microw. Wirel. Compon. Lett.* **2014**, *24*, 56–58. [CrossRef]
25. Rücker, H.; Heinemann, B.; Fox, A. Half-Terahertz SiGe BiCMOS Technology. In Proceedings of the IEEE 12th Topical Meetings on Silicon Monolithic Integrated Circuits in RF Systems (SiRF), Santa Clara, CA, USA, 16–18 January 2012; pp. 133–136.
26. Edwards, M.L.; Sinsky, J.H. A new criterion for linear 2-port stability using a single geometrically derived parameter. *IEEE Trans. Microw. Theory Tech.* **1992**, *40*, 2303–2311. [CrossRef]

Article

Tin Dioxide Thin Film with UV-enhanced Acetone Detection in Microwave Frequency Range

Artur Rydosz *, Kamil Staszek, Andrzej Brudnik and Slawomir Gruszczynski

Department of Electronics, AGH University of Science and Technology, 30059 Krakow, Poland
* Correspondence: artur.rydosz@agh.edu.pl; Tel.: +48-126-172-594

Received: 29 July 2019; Accepted: 29 August 2019; Published: 30 August 2019

Abstract: In this paper, the UV illumination effect for microwave gas sensors based on the tin dioxide was verified. A UV LED with emission wavelength close to the absorption edge of the SnO_2 gas-sensing layer was selected as the UV source. The developed gas sensors were tested under exposure to acetone in the 0–200 ppm range at room temperature. The sensor's complex reflection coefficient corresponding to target gas concentration was measured with the use of a five-port reflectometer system exhibiting enhanced uncertainty distribution, which allows for the detection of low gas concentration. The UV illumination significantly emphasizes the sensors' response in terms of both magnitude and phase for low gas concentrations, in contrast to previously reported results, in which only the reflection coefficient's phase was affected. The highest responses were obtained for modulated UV illumination.

Keywords: gas sensors; acetone detection; microwave application; UV illumination

---

1. Introduction

Acetone (C_2H_5OH) is a colorless, mobile, flammable liquid that serves as an important solvent in chemistry and industry. Recently, it has become attractive for biomedical applications, where it is considered a biomarker of diabetes, due to its presence in exhaled breath in various concentrations for healthy and diabetic patients [1–5]. Patients with diabetes tend to have higher acetone levels (1.25–2.5 ppm) in their breath than healthy people (0.2–0.8 ppm). On the other hand, the inhalation of acetone at higher concentrations (200–2000 ppm) may lead to hepatotoxic effects, causing liver damage [6]. Therefore, it is essential to detect acetone at lower and higher concentrations. The detection at higher concentrations is covered by commercially available sensors, e.g., TGS Figaro [7], but detectors able to detect acetone in the sub-ppm range are still under investigation. In the last few years, a number of papers have focused on enhanced acetone detection, utilizing various methods, such as optical detection [8–10], electrochemical sensors [11–13], metal oxides (MOXs)-based sensors [14–20], and analytical systems [21–25]. Microwave-based gas sensors with various gas-sensitive layers, including organic layers [26,27] and various MOXs layers [28–30], were investigated by the authors as well. The recently obtained results have shown that microwave gas sensors based on metal oxides can easily be utilized for acetone detection in the ppm range at room temperature [28–30]. However, for such sensors, the response/recovery time(s) at room temperature is longer than for conventional applications, where the operating temperature is usually in the 300–500 °C range. To overcome this limitation and to increase the gas sensor response, a UV illumination can be utilized.

Recently, Khan et al. [31] presented the development of a toluene detector based on deep UV absorption spectrophotometry. The setup was tested for different toluene concentrations (10–100 ppm) and a linear relationship between gas concentration and the absorbance was observed. The sensitivity and selectivity of the setup can be improved by coupling it with a preconcentration unit, e.g., [32–34]. Bastates et al. [35] measured the effect of UV illumination (369 nm, 17 mW) on the detection of ammonium nitrate (NH_4NO_3) by a ZnO-coated nanospring sensor operated at room temperature [35]. The investigation results showed that UV illumination reduces surface band bending and reduces the sensor recovery time between detection events by shortening the decay time of the signal [35]. A good review of light-activated metal oxide gas sensors was delivered by Xu and Ho [36]. The authors reviewed the progress of light-activated conductometric gas sensors based on metal oxides, such as pure metal oxides, 1D nanostructures, and porous nanostructures. They confirmed that light intensity at a specific wavelength can change the sensing response and even tune the device's selectivity, however, the correct intensity level, as well as film thickness, must be investigated for each MOX separately [36].

In this paper, a microwave gas sensor based on the tin dioxide (SnO_2) gas layer with UV illumination is investigated for acetone detection in the 0–200 ppm range at room temperature. The gas-sensing properties of SnO_2 at microwave frequencies were previously confirmed and presented in [30]. The obtained results strongly confirm that the UV illumination increases the developed sensor's sensitivity for acetone, allowing for the detection of lower gas concentrations. Moreover, a linear relationship between acetone concentration and magnitude/phase changes of the sensor's reflection coefficient was observed.

2. Materials and Methods

2.1. SnO_2 Deposition Technology

Tin dioxide (SnO_2) thin films were deposited in RF (radio frequency, 13.56 MHz) mode from the Sn metallic target by applying reactive sputtering under a mixture of 80% argon and 20% oxygen and by applying GLAD (glacing angle deposition). The base vacuum and deposition vacuum were 1×10^{-5} mbar and 2×10^{-2} mbar, respectively. The deposition temperature was set to 200 °C and deposition time was adjusted to deposit various thicknesses (50 nm, 250 nm, and 500 nm), with a constant power of 50 W. The films' thicknesses were measured post-process using a TalyStep profilometer (Taylor Hobson, Leicester, UK). The fabricated sensors were tested for gas-sensing applications and the highest responses were obtained for 250 nm thin films [30], therefore these sensors were investigated for the UV illumination effect. The sputtering deposition system was presented previously in [30].

2.2. Microwave Measurements

As was shown in [30], the above described SnO_2 layer changes its permittivity in the microwave frequency range when exposed to acetone. This phenomenon has been used for indirect acetone concentration measurement with the use of the microwave measurement system reported in [37], operating at a frequency of 2.4 GHz. This system is composed of a microwave sensor with the SnO_2 as a gas-sensing layer and a dedicated five-port reflectometer for measuring the sensor's response.

The microwave sensor is presented in Figure 1a. It was composed of two baluns, between which a coupled-line section covered with a SnO_2 layer was inserted. These baluns ensured the odd-mode excitation of the mentioned coupled-line section, enhancing the sensor's sensitivity. When exposed to acetone present in the cavity, the SnO_2 layer changed its permittivity, affecting the electromagnetic field distribution along the coupled-line section, which in turn changed the sensor's transmission coefficient. For further sensitivity increase, the sensor was used in a single-port configuration with port #2 left opened. In such a case the reflection coefficient seen at port #1 (the measured value) is approximately equal to the transmission coefficient's squared, since a microwave signal propagates through the sensor from port #1 to port #2 and is reflected. As a result, such a sensor configuration doubles the impact of the permittivity change on the measured reflection coefficient. To obtain good measurement quality, the reflection coefficient measurement was realized with the use of the recently developed five-port reflectometer [37]. It exhibited significantly enhanced measurement uncertainty for the reflection coefficient's range, corresponding to the utilized sensor's reflection coefficient (magnitude equal to 0.8 ± 0.2 and phase equal to $180° \pm 15°$) [37], with respect to the classic reflectometers optimized for all reflection coefficients (magnitude not exceeding 1, arbitrary phase). It consisted of a five-port passive power distribution network, signal source, and three microwave power detectors, the readings of which were translated to the measured complex reflection coefficient. Since the utilized system was able to measure both the magnitude and phase of the sensor's reflection coefficient, the gas-sensor response can be defined twofold, i.e., as magnitude difference $\Delta|s_{11}|$ and phase difference $\Delta arg[s_{11}]$ of the reflection coefficients measured under exposure to target gas $s_{11\ gas}$ and air $s_{11\ air}$:

$$\Delta|s_{11}| = \left|s_{11\ gas}\right| - |s_{11\ air}|; \tag{1}$$

$$\Delta arg[s_{11}] = arg\left[\frac{s_{11\ gas}}{s_{11\ air}}\right]. \tag{2}$$

The entire measurement system is shown in Figure 1c.

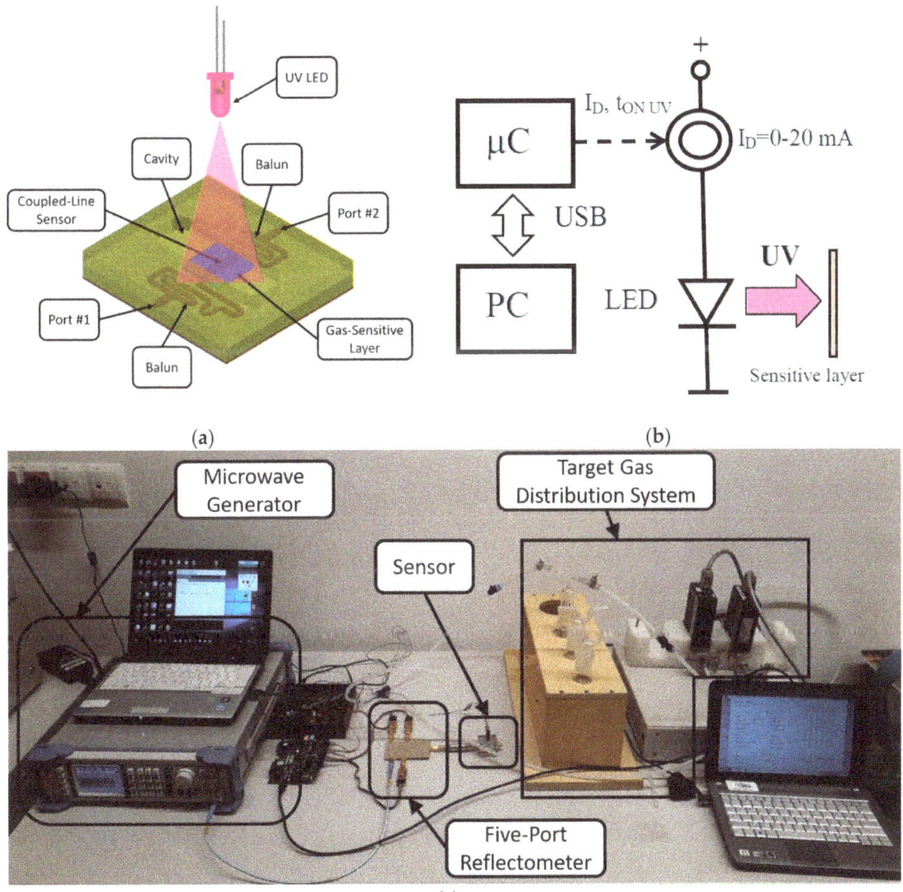

Figure 1. The microwave gas sensor system: (**a**) sketch of the microwave gas sensor; (**b**) sketch of the UV illumination system; (**c**) photo of the gas-sensing system.

2.3. Gas-sensing Protocol with UV

Figure 1b shows the UV LED supply circuit. The UV source was a UV LED (OSV2YL5111A) with $\lambda = 375$ nm. Figure 2 shows the transmission of the SnO_2 thin film working as the gas-sensitive layer and emission of the UV LED. The transmission was measured by a Lambda 19 Spectrophotometer (Perkin-Elmer, UK) and the UV LED emission by a monochromator SPM-2 (Carl-Zeiss, Jena, Germany). The λ (375 nm) was chosen to be in the absorption edge of the gas-sensitive layer, as it was presented in Figure 2.

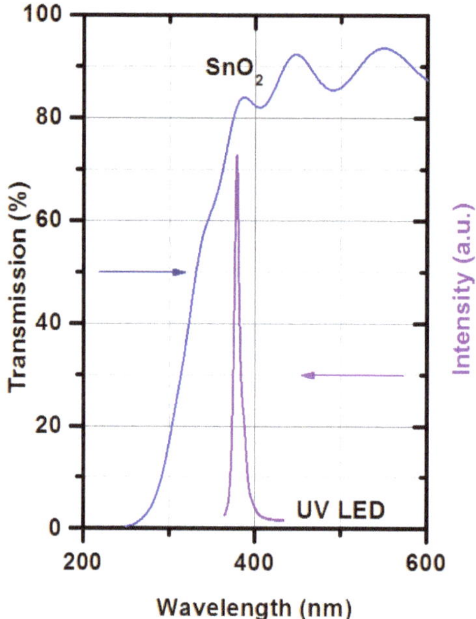

Figure 2. The SnO_2 transmission with the intensity of the UV LED diode (OSV2YL5111A) in the function of wavelength.

To control the diode current ID (0–20 mA), a microcontroller Atmega 328 was used with a dedicated software on PC via a USB port. The measurements were conducted in two modes: CW (continuous wave) and PWM (pulse width modulation). The gas-dosing system was based on the mass flow controllers 1179B (MKS Instruments, New York, NY, USA), Dreschel bottles, and gas canisters (Air Liquid, Krakow, Poland). The data was collected by dedicated software and further processed by developed algorithms. All experiments were conducted at room temperature and 50% relative humidity level, controlled by an AC system in the laboratory.

3. Results and Discussion

3.1. Gas-sensing Characteristics for Continouse UV Radiation

Figure 3 shows the gas-sensing characteristics under exposure to 0–200 ppm acetone at continuous UV illumination with various LED diode currents. The developed system measured both the phase and magnitude of the sensor's reflection coefficient. Figure 3a,c show the phase and magnitude changes under exposure to 0–200 ppm of acetone, respectively. The lowest acetone concentration used during the measurements was 20 ppm, based on the limitation of the gas distribution system, for which the response was 0.02 of the magnitude change and 1° of phase change (at 10 UV CW illumination). On the other hand, the magnitude and phase noise level were equal to 0.0033 and 0.21°, respectively (calculated as 3σ values). As can be observed, the obtained values were significantly larger than the noise level for both magnitude and phase, hence a lower concentration of acetone than 20 ppm could be measured. Nevertheless, for such low concentrations, the long-term drift seen in both measured magnitude and phase responses becomes dominant and needs to be cancelled to obtain more accurate results. The measurement uncertainty was below 5%, and error bars are represented by size of the measurement points.

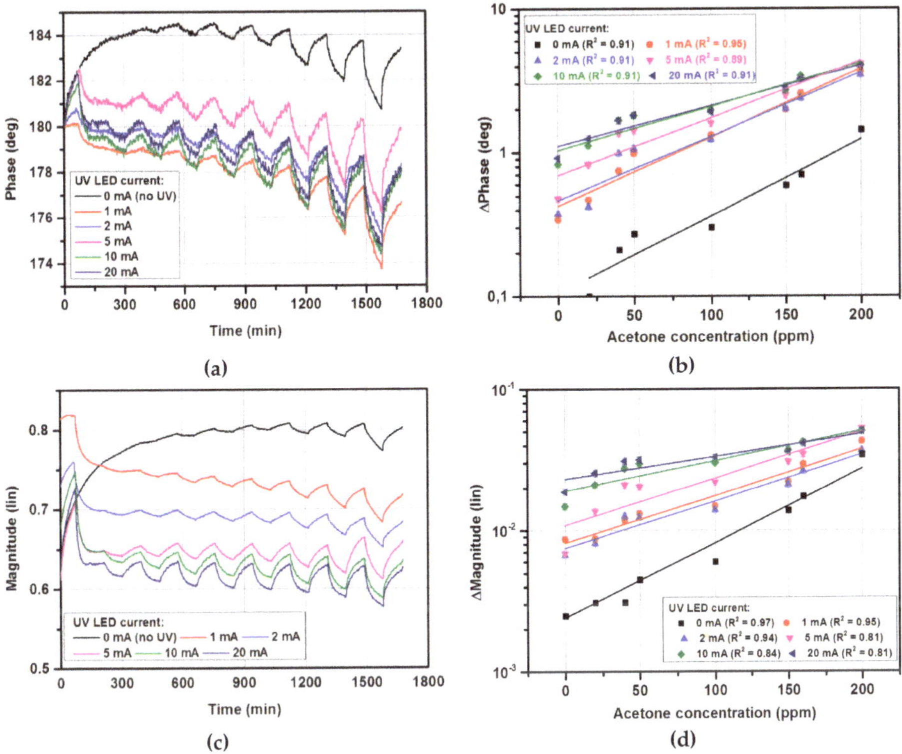

Figure 3. The gas-sensing characteristics under exposure to acetone in the 0–200 ppm range: (**a**) phase changes in time domain; (**b**) calibration curve of phase changes; (**c**) magnitude changes in time domain; (**d**) calibration curve of magnitude changes.

As can be observed, the UV illumination significantly emphasized the sensor response in terms of both magnitude and phase for low gas concentrations, in contrast to previously reported results, in which only the reflection coefficient's phase was affected [37]. However, this effect decreased for higher concentrations, for which an emphasis was not required. Hence, UV illumination increased the sensor's sensitivity and simultaneously held the measured reflection coefficient in the area described in Section 2.2, for which the utilized five-port reflectometer was optimized. Thanks to this effect, low measurement uncertainty was preserved for all the measured gas concentrations. As seen in Figure 3b,d, the magnitude changes were of higher quality than the phase changes and could be easily used as gas sensor response signal, defined by (2) in Section 2.2. Increasing the diode current increased the gas sensor's response, however, the UV LED current higher than 10 mA did not provide further sensitivity enhancement, therefore 10 mA was set for further experiments, to restrain the power consumption. Furthermore, Figure 2b shows that applying UV illumination even with a 1 mA diode current significantly improved the gas sensor's response in comparison with no UV illumination.

3.2. Gas-sensing Characteristics for Modulated UV Illumination

After measurement with continuous wave (CW) UV illumination at various UV currents (Section 3.1), the developed sensor was tested under exposure to acetone and three various periods of UV switching were applied. The periods were T_1, T_2, and T_3, where the UV light was switched ON/OFF (current amplitude was 10 mA) for 20 s, 200 s, and 2000 s, respectively. Figure 4 presents the obtained results. As can be observed, the magnitude and phase changes for UV modulation with T_1 and T_2 periods had lower values in comparison with 5 mA continuous wave illumination. The modulation with the 2000 s period (T_3) affected the magnitude measurements, where the switching effect is visible. The switching effect with T_1/T_2 periods was unnoticeable. On the other hand, T_3 was ~1/3 of the exposure time (5400 s), hence, for a single acetone concentration, the UV light was switched ON/OFF around three times (Figure 4a). Therefore, to calculate the sensor's sensitivity using (1), the min/max values must be taken considering three ripples, in contrast to a single ripple occurring for every other case. A reduction in UV illumination cycles number responded with higher gas sensor responses (Figure 4c, dotted line), calculated from the magnitude signal's envelope. Moreover, the limit of detection for sensors with UV signal modulation close to the exposure time was higher than for continuous wave modulation.

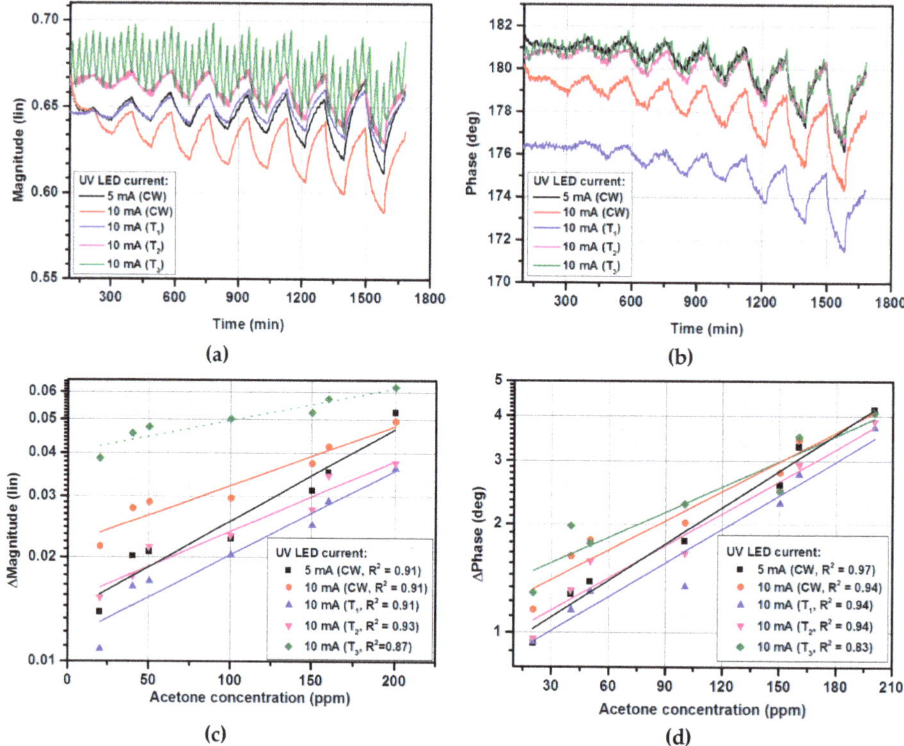

Figure 4. The gas-sensing characteristics under exposure to acetone in the 0–200 ppm range: (**a**) magnitude changes in time; (**b**) phase changes in time; (**c**) calibration curve for magnitude changes; (**d**) calibration curve for phase changes.

3.3. Gas-sensing Characteristics for Pulse Width Modulated (PWM) UV Illumination

The developed sensors were tested under exposure to acetone and pulse width modulation (PWM) with various duty cycles: 30%, 50%, and 70%, and 10 mA amplitude. As can be observed in Figure 5, the duty cycles correspond to continuous waves with 3 mA, 5 mA, and 7 mA, respectively. It has to be underlined that the period of duty cycles was equal to T_1 from Section 2.1. The magnitude/phase measurements exhibited slight drift, which is most likely due to local temperature and humidity changes and synthetic gas impurities. The total time for one series of measurements was 28 h.

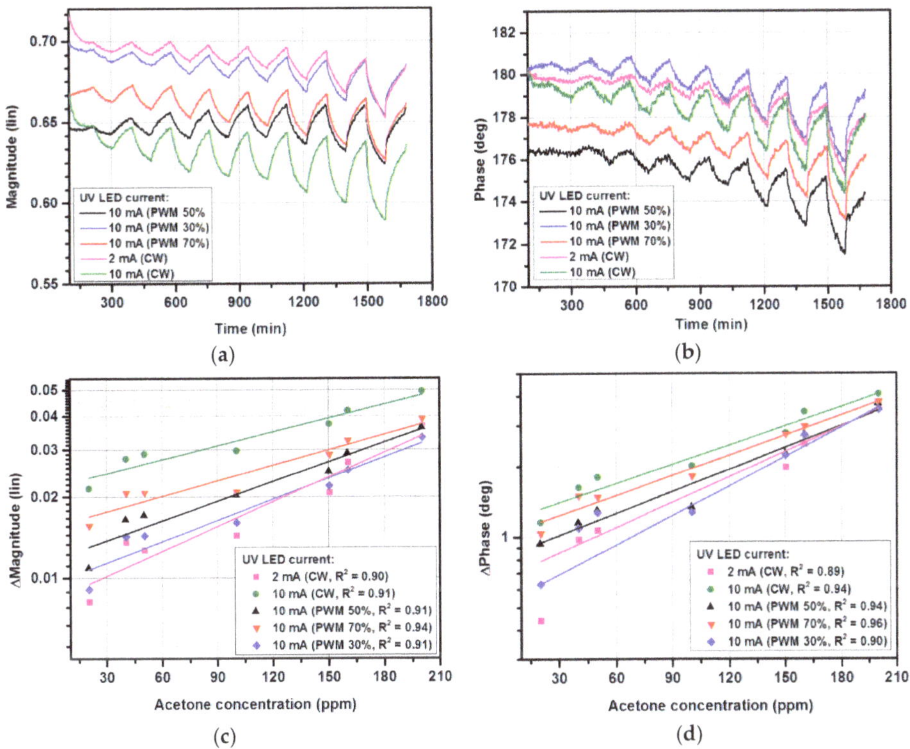

Figure 5. The gas-sensing characteristics under exposure to acetone in the 0–200 ppm range: (**a**) magnitude changes in time; (**b**) phase changes in time; (**c**) calibration curve for magnitude changes; (**d**) calibration curve for phase changes.

Figure 6 shows two UV illumination sequences (T_4, T_5) with a gas-sampling sequence. The gas-sampling process started with 30 min without target gas (only synthetic air), then 90 min periods with ON/OFF gas were set. The various gas concentrations were used as presented in Figure 4. The UV illumination switching was preset to start with 'high' UV signal for 10 min. and then 'low' for 20 min, and 30 min without UV light (sequence T_4), or to start with 'low', 'high', and without (sequence T_5). The idea behind using various PWM sequences instead of CW signals was to reduce power consumption, and to introduce the UV at the beginning and at the end of exposure to gas and in the middle of the no-gas period. Such a method was proposed to verify the influence of UV illumination for gas-sensing characteristics. In sequence T_5, the UV light was applied with high signal (10 mA) at the same time as when the sensor was exposed to the target gas, which should increase the sensitivity, and the low signal (2 mA) was kept, to stabilize the response at lower power consumption. The same method was repeated at the end of the gas-sampling procedure. After dosing the target gas, the synthetic air was introduced to the gas chamber without UV illumination, which should react with a longer recovery time. To verify this hypothesis, the UV illumination was again switched on in the middle of exposure to synthetic air. The sequence was repeated for all target gas concentrations. Sequence T_5 was prepared in reverse to T_4, where the low signal was first applied and then a high signal. As can be observed, the magnitude (Figure 7a) and phase (Figure 7b) were higher in comparison with the 2 mA and 10 mA continuous wave, however, the sensor response defined as magnitude/phase changes did not change significantly. Further investigations are needed to find the optimal working conditions, where a continuous UV signal can be successfully replaced for modulated ones and where not only the UV LED current amplitude will be modulated but also the wavelength of the UV source.

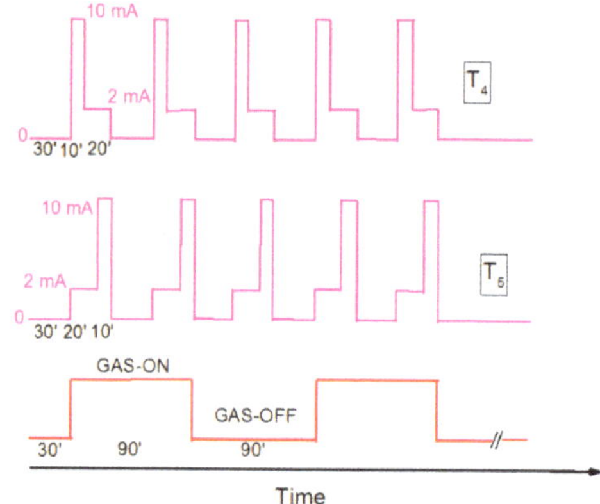

Figure 6. The PWM sequences named T_4 and T_5, and gas-sampling sequence.

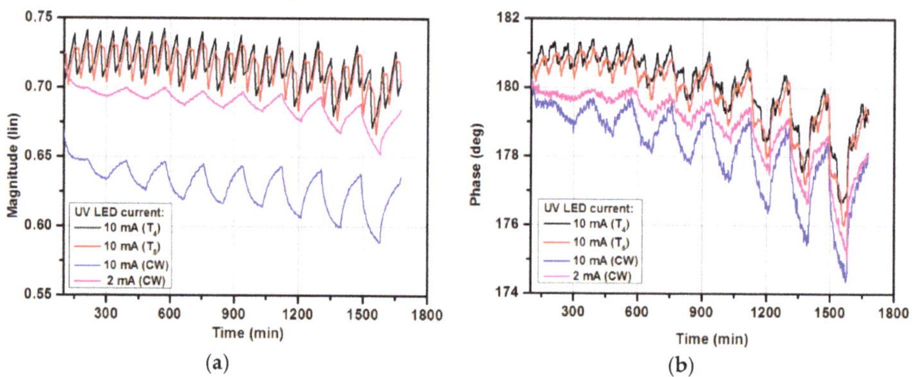

Figure 7. The gas-sensing characteristics under exposure to acetone in the 0–200 ppm range: (**a**) magnitude changes with various PWM cycles (T_4, T_5); (**b**) phase changes with various PWM cycles (T_4, T_5).

4. Conclusions

Microwave gas sensors operated at microwave frequencies, such as 2.4 GHz and at room temperature, can be applied in many various applications. Previously, various metal oxides have been proposed as a gas-sensitive layer for acetone detection [30]. The conducted research has confirmed that the sensitivity of the microwave gas sensors based on SnO_2 can be further improved by applying UV illumination, which emphasizes the sensor's response to lower gas concentrations. Simultaneously, it does not magnify this response for higher concentrations, preserving the measured reflection coefficient in the range for which the measurement uncertainty of the utilized system is optimized. As a result, a good measurement quality has been obtained for a wide range of gas concentrations. Various experimental conditions were tested. The highest sensitivity was obtained for a UV (375 nm) current of 10 mA at continuous wave. However, further investigation is needed to find the optimal conditions with PWM modulation and to select the UV diode wavelength in terms of various metal oxide layers.

Author Contributions: Conceptualization A.R., K.S., A.B., S.G.; methodology A.R., K.S., A.B.; software K.S. and A.B.; validation A.R..; investigation A.R., K.S., A.B.; writing—original draft preparation A.R., K.S., A.B., S.G.; writing—review and editing A.R., K.S., A.B., S.G.; visualization A.R., K.S., A.B., S.G.; project administration A.R., K.S., S.G.

Funding: This research was partially funded by the National Science Centre, Poland 2017/26/D/ST7/00355 and by the National Science Centre, Poland 2016/23/D/ST7/00481, and by the National Science Centre, Poland 2016/22/E/ST7/00021.

Conflicts of Interest: The authors declare no conflict of interest.

References

1. Crofford, O.B.; Mallard, R.E.; Winton, R.E.; Rogers, N.L.; Jackson, J.C.; Keller, U. Acetone in breath and blood. Trans. *Am. Clin. Climatol. Assoc.* **1977**, *88*, 128–139.
2. Minh, T.D.C.; Blake, D.R.; Galassetti, P.R. The clinical potential of exhaled breath analysis for diabetes mellitus. *J. Diabetes Res.* **2012**, *97*, 195–205. [CrossRef] [PubMed]
3. Amann, A.; Smith, D. *Volatile Biomarkers. Non-Invasive Diagnosis in Physiology and Medicine*; Elsevier: Amsterdam, The Netherlands, 2013.
4. Afreen, S.; Zhu, J.-J. Rethinking EBAD: Evolution of smart noninvasive detection of diabetes. *TrAC Trends Anal. Chem.* **2019**, *118*, 477–487. [CrossRef]
5. Rydosz, A. Sensors for Enhanced Detection of Acetone as a Potential Tool for Noninvasive Diabetes Monitoring. *Sensors* **2018**, *18*, 2298. [CrossRef] [PubMed]
6. Available online: http://www.hmdb.ca/metabolites/HMDB0001659 (accessed on 26 July 2019).
7. Available online: http://www.figarosensor.com/products/822pdf.pdf (accessed on 26 July 2019).
8. Li, J.; Smeeton, T.M.; Zanola, M.; Barrett, J.; Berryman-Bousquet, V. A compact breath acetone analyzer based on an ultraviolet light emitting diode. *Sens. Actuators B* **2018**, *273*, 76–82. [CrossRef]
9. Das, P.; Ganguly, S.; Mondal, S.; Bose, M.; Das, A.K.; Baneerjee, S.; Fas, N.C. Heteroatom doped photoluminescent carbon dots for sensitive detection of acetone in human fluids. *Sens. Actuators B* **2018**, *266*, 583–593. [CrossRef]
10. Chen, Y.; Owyeung, R.E.; Sonkusale, S.R. Combined optical and electronic paper-nose for detection of volatile gases. *Anal. Chim. Acta* **2018**, *1034*, 128–136. [CrossRef] [PubMed]
11. Hao, X.; Wu, D.; Wang, Y.; Ouyang, J.; Wang, J.; Liu, T.; Liang, X.; Zhang, C.; Liu, F.; Yan, X.; et al. Gas sniffer (YSZ-based electrochemical gas phase sensor) toward acetone detection. *Sens. Actuators B* **2019**, *278*, 1–7. [CrossRef]
12. Liu, X.; Zhao, K.; Sun, X.; Duan, X.; Zhang, C.; Xu, X. Electrochemical sensor to environmental pollutant of acetone based on Pd-loaded on mesoporous In_2O_3 architecture. *Sens. Actuators B* **2019**, *290*, 217–225. [CrossRef]
13. Liu, F.; Ma, C.; Hao, X.; Yang, C.h.; Zhu, H.; Liang, X.; Sun, P.; Liu, F.; Chuai, X.; Lu, G. Highly sensitive gas sensor based on stabilized zirconia and $CdMoO_4$ sensing electrode for detection of acetone. *Sens. Actuators B* **2017**, *248*, 9–18. [CrossRef]
14. Yang, F.; Wang, F.; Guo, Z. Characteristics of binary WO_3@CuO and ternary WO_3@PDA@CuO based on impressive sensing acetone odor. *J. Colloid Interface Sci.* **2018**, *524*, 32–41. [CrossRef] [PubMed]
15. Righettoni, M.; Tricoli, A.; Pratsinis, S.E. Si: WO_3 Sensors for highly selective detection of acetone for easy diagnosis of diabetes by breathe analysis. *Anal. Chem.* **2010**, *82*, 3581–3587. [CrossRef] [PubMed]
16. Szkudlarek, A.; Kollbek, K.; Klejna, S.; Rydosz, A. Electronic sensitization of CuO thin films by Cr-doping for enhanced gas sensor response at low detection limit. *Mater. Res. Express* **2018**, *5*, 126406. [CrossRef]
17. Rydosz, A.; Szkudlarek, A.; Ziabka, M.; Domanski, K.; Maziarz, W.; Pisarkiewicz, T. Performance of Si-Doped WO_3 Thin Films for Acetone Sensing Prepared by Glancing Angle DC Magnetron Sputtering. *IEEE Sens. J.* **2016**, *16*, 1004–1012. [CrossRef]
18. Hu, J.; Yang, J.; Wang, W.; Xue, Y.; Sun, Y.; Li PLian, K.; Zhang, W.; Chen, L.; Shi, J.; Chen, Y. Synthesis and gas sensing properties of NiO/SnO_2 herarchical structures toward ppb-level acetone detection. *Mater. Res. Bull.* **2018**, *102*, 294–303. [CrossRef]

19. Wang, X.-F.; Ma, W.; Jiang, F.; Cao, E.-S.; Sun, K.-M.; Cheng, L.; Song, X.-Z. Prussian Blue analogue derived prorous $NiFe_2O_4$ nanocubes for low-concentration acetone sensing at low working temperature. *Chem. Eng. J.* **2018**, *338*, 504–512. [CrossRef]
20. Asgari, M.; Saboor, F.H.; Mortazavi, Y.; Khodadadi, A.K. SnO_2 decorated SiO_2 chemical sensors: Enhanced sensing performance toward ethanol and acetone. *Mater. Sci. Semicond. Process.* **2018**, *68*, 87–96. [CrossRef]
21. Deng, C.; Zhang, J.; Yu, X.; Zhang, W.; Zhang, X. Determination of acetone in human breath by gas chromatography-mass spectrometry and solid-phase microextraction with on-fiber derivatization. *J. Chromatogr. B* **2004**, *810*, 269–275. [CrossRef]
22. Pauling, L.; Robinson, A.B.; Teranishi, R.; Cary, P. Quantitative analysis of urine vapor and breath by gas-liquid partition chromatography. *Proc. Natl. Acad. Sci. USA* **1971**, *68*, 2374–2376. [CrossRef]
23. Righettoni, M.; Schmid, A.; Amann, A.; Pratsinis, S.E. Correlations between blood glucose and breath components from portable gas sensors and PTR-TOF-MS. *Breath Res.* **2013**, *7*, 037110. [CrossRef]
24. Saidi, T.; Zaim, O.; Moufid, M.; Bari, N.E.; Ionescu, R.; Bouchikhi, B. Exhaled breath analysis using electronic nose and gas chromatography-mass spectrometry for non-invasive diagnosis of chronic kidney disease, diabetes mellitus and healthy subjects. *Sens. Actuators B* **2018**, *257*, 178–188. [CrossRef]
25. Szymanska, E.; Tinnevelt, G.H.; Brodrick, E.; Williams, M.; Davies, A.N.; van Manen, H.-J.; Buydens, L.M.C. Increasing conclusiveness of clinical breath analysis by improved baseline correction of multi capillary column - ion mobility spectrometry (MCC-IMS) data. *J. Pharm. Biomed. Anal.* **2016**, *127*, 170175. [CrossRef] [PubMed]
26. Rydosz, A.; Maciak, E.; Wincza, K.; Gruszczynski, S. Microwave-based sensors with phthalocyanine films for acetone, ethanol and methanol detection. *Sens. Actuators B* **2016**, *237*, 876–886. [CrossRef]
27. Staszek, K.; Rydosz, A.; Maciak, E.; Wincza, K.; Gruszczynski, S. Six-port microwave system for volatile organic compounds detection. *Sens. Actuators B* **2017**, *245*, 882–894. [CrossRef]
28. Rydosz, A. The use of copper oxide thin tilms in gas-sensing applications. *Coatings* **2018**, *8*, 425. [CrossRef]
29. Rydosz, A.; Maciak, E.; Wincza, K.; Slawomir, G. Microwave-based sensors for exhaled acetone and ethanol detection. In Proceedings of the 2018 International Conference on Electromagnetics in Advanced Applications (ICEAA), Cartagena de Indias, Colombia, 10–14 September 2018. [CrossRef]
30. Rydosz, A.; Brudnik, A.; Staszek, K. Metal Oxide Thin Films Prepared by Magnetron Sputtering Technology for Volatile Organic Compound Detection in the Microwave Frequency Range. *Materials* **2019**, *12*, 877. [CrossRef] [PubMed]
31. Khan, S.; Newport, D.; Le Calve, S. Development of a Toluene Detector Based on Deep UV Absorption Spectrophotometry Using Glass and Aluminum Capillary Tube Gas Cells with a LED Source. *Micromachines* **2019**, *10*, 193. [CrossRef] [PubMed]
32. Rydosz, A.; Maziarz, W.; Pisarkiewicz, T.; Domanski, K.; Grabiec, P. A gas micropreconcentrator for low level acetone measurements. *Microelectron. Reliab.* **2012**, *52*, 2640–2646. [CrossRef]
33. Rydosz, A. Micropreconcentrators in silicon-glass technology for the detection of diabetes biomarkers. *J. Microelectron. Electron. Compon. Mater.* **2014**, *44*, 126–136.
34. Rydosz, A.; Maziarz, W.; Pisarkiewicz, T.; Bartch de Torres, H.; Mueller, J. Amicropreconcentrator design using low temperature cofired ceramics technology for acetone detection applications. *IEEE Sens. J.* **2013**, *13*, 1889–1896. [CrossRef]
35. Bastatas, L.D.; Wagle, P.; Echeverria, E.; Austin, A.J.; McIlroy, D.N. The Effect of UV Illumination on the room temperature detection of vaporized ammonium nitrate by a ZnO coated nanospring-based sensors. *Materials* **2019**, *12*, 302. [CrossRef] [PubMed]
36. Xu, F.; Ho, H.P. Light-activated metal oxide gas sensors: A review. *Micromachines* **2017**, *8*, 333. [CrossRef] [PubMed]
37. Staszek, K.; Szkudlarek, A.; Kawa, M.; Rydosz, A. Microwave system with sensor utilizing GO-based gas-sensitive layer and its application to acetone detection. *Sens. Actuators B* **2019**, *297*, 126699. [CrossRef]

Article

Experimental Microwave Complex Conductivity Extraction of Vertically Aligned MWCNT Bundles for Microwave Subwavelength Antenna Design

Charlotte Tripon-Canseliet [1,*], Stephane Xavier [2], Yifeng Fu [3], Jean-Paul Martinaud [4], Afshin Ziaei [2] and Jean Chazelas [4]

1 Physics and Material Science Laboratory, Sorbonne Universités, CNRS-ESPCI, 75005 Paris, France
2 THALES Research and Technology, 91767 Palaiseau, France
3 Electronics Materials and Systems Laboratory, Department of Microtechnology and Nanoscience, Chalmers University of Technology, SE - 412 96 Gothenburg, Sweden
4 THALES Defence Mission Systems, 78 851 Elancourt, France
* Correspondence: charlotte.tripon-canseliet@espci.fr; Tel.: +33-664-545-363

Received: 10 May 2019; Accepted: 12 August 2019; Published: 27 August 2019

Abstract: This paper reports the extraction of electrical impedance at microwave frequencies of vertically aligned multi-wall carbon nanotubes (VA MWCNT) bundles/forests grown on a silicon substrate. Dedicated resonating devices were designed for antenna application, operating around 10 GHz and benefiting from natural inductive/capacitive behavior or complex conductivity in the microwave domain. As obtained from S-parameters measurements, the capacitive and inductive behaviors of VA MWCNT bundles were deduced from device frequency resonance shift.

Keywords: multi-wall carbon nanotubes; microwave impedance; small antennas

---

1. Introduction

Carbon nanotubes (CNTs) have been extensively studied over the last decades due to their exceptional electrical, thermal, and mechanical properties. In addition to these properties, applications of CNTs in the microwave domain have also appeared, as numerous research works have been devoted to the elaboration of active and passive radio-frequency (RF) devices such as resonators and field-effect transistors (FETs) [1,2], but also to the development of chemical and mechanical sensors [3–5] for environment monitoring and biomedical applications. Several technical approaches are competing for microwave antenna miniaturization prospects, as the resolution of numerous technological processes is converging drastically down to the nanometer scale. At the same time, hybrid material compatibility can now be implemented as high-resolution characterization tools such as near-field probe techniques coupled to large broadband vector network analysis become more affordable [6–9].

Major challenges remain in the qualification of this material in the microwave domain in terms of complex impedance or conductivity. In this research field, 1D and 2D materials have recently validated their eligibility of implementation in microwave circuit design, reaching a high level of innovation, as confirmed from theory [10,11]. Indeed, graphene and metallic CNT material conductivity, offering a non-negligible imaginary part, now stands as the best candidate for antenna miniaturization (Table 1) following the initial experimental validations [12]. These technical approaches are competing with surface-mode propagation solutions superimposed by metallic/dielectric interfaces. Plasmonic structures are the most studied nanoscale configuration in the THz regime [13].

Table 1. Comparison of lineic electrical properties of a 10-nm-diameter cylinder.

Parameter	Perfect Electrical Conductor (PEC) Wire	Individual MWCNT (Multi-Wall Carbon Nanotube)
Resistance	5 mΩ/µm	7 kΩ/µm
Inductance	1 pH/µm	20 nH/µm
Capacitance	5 aF/µm	0.5 fF/µm

At microwave frequencies, metasurface circular topologies leading to a leaky-wave radiation such as in [14,15] allow a maximum miniaturization factor of λ/50 for each individual element. CNT-based composite materials technology in which multi-wall (MW) or single-wall (SW) CNTs are dispersed randomly in a solution or polymeric layer have demonstrated their efficiency in low-frequency antenna design on flexible substrates [16].

At present, the identification of electromagnetic properties at RF/microwave frequencies has become a milestone in order to understand and design new components for future implementation in the next generation of CNT-based microwave systems.

In this paper, we report on the electromagnetic material properties of vertically aligned CNT bundles grown on a substrate by the exploitation of a de-embedding technique developed for integrated technology. Using this approach, the experimental complex conductivity of MWCNTs bundles processed from well-controlled vertical growth technology in the microwave domain is presented from impedance extraction obtained from on-wafer S-parameters measurements.

2. MWCNT Material Properties from Theory

An individual single-wall carbon nanotube (SWCNT) consists of a spatially unique angled rolling of a graphene sheet that depends on the associated chiral vector, which defines its degree of metallicity.

Frequency-dependent graphene conductivity can be approximated using the Kubo formula (1), with the chemical potential μ_c and the relaxation rate Γ. This complex expression, in which Γ was experimentally estimated to be 0.1 meV, exhibits a complex behavior with σ_{real} and σ_{imag} as real and imaginary parts of $\sigma_s(\omega)$ depending on the value of μ_c. From a dyadic Green's function formulation of electromagnetic field propagation through Sommerfeld integrals, Transverse Electric (TE) or Transverse Magnetic (TM)-mode surface wave propagation operations have been validated for positive or negative values of σ_{imag} from [3].

Starting from the graphene conductivity formulation (1), a specific SW rolling configuration leading to a metallic behavior also implies a complex conductivity definition as in a chiral case (2). A MW formation of CNTs has to be selected in order to ensure a metallic behavior from catalyst atom choice and growth technique. From the literature, an RLC circuit representation study of individual SW or MW CNTs from electron gas theory states a high linear resistance, static capacitance, and kinetic inductance per micrometer values at several orders of magnitude higher than a standard metallic gold wire with the same dimensions (i.e., with 5 nm radius and 300 µm distance from a ground plane; Table 1).

$$\sigma_s(\omega) = i\frac{1}{\pi\hbar^2}\frac{e^{2k_BT}}{\omega + i2\Gamma}\left\{\frac{\mu_c}{k_BT} + 2\ \ln\left[\exp\left(-\frac{\mu_c}{k_BT}\right)+1\right]\right\} + i\frac{e^2}{4\pi\hbar}\ln\left[\frac{2|\mu_c| - \hbar(\omega + i2\Gamma)}{2|\mu_c| + \hbar(\omega + i2\Gamma)}\right] \quad (1)$$

$$\sigma_{zz_{chiral}} = -j\frac{8\pi e^2\gamma_0\sqrt{3}}{h^2\sqrt{m^2 + n^2(\omega - j\nu) + mn}} \text{ with } 2n + m = N \quad (2)$$

$$\sigma_{bundle} = \sigma_{b0} + j\sigma_b(\omega) \quad (3)$$

As a consequence, from these statements, the existence of an imaginary character of electrical conductivity (i.e., an inductive or capacitive behavior in frequency) presents strong interest in electronic circuits designed from the material itself. Furthermore, considering a vertically aligned collective representation of individual CNTs as in a bundle, a complex expression of bundle-equivalent conductivity must be assumed as in (3), with a static parameter σ_{b0} and a frequency-dependent imaginary part $\sigma_b(\omega)$.

3. Microwave Material Parameters Identification Procedure

3.1. Microwave Structures Design

Preliminary resonant structures in coplanar waveguide (CPW) technology with suitable taper-based electrodes (Figure 1) were designed using commercial 3D electromagnetic simulation code (Ansys-HFSS) with technological process implementation.

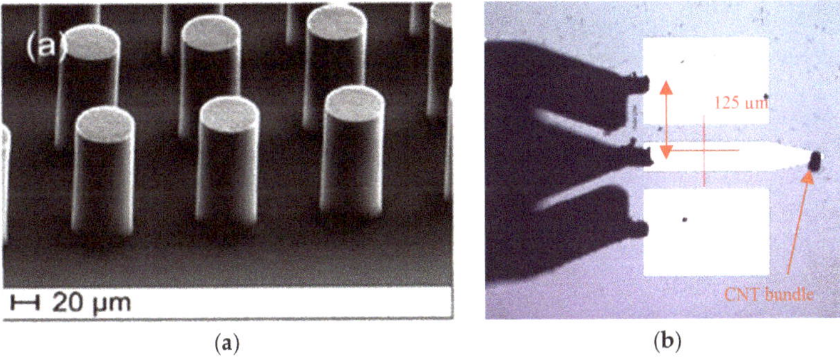

(**a**) (**b**)

Figure 1. Technological process and Characterization of the vertically aligned multi-wall carbon nanotubes (MWCNTs)-based device. (**a**) SEM (scanning electron microscope) image of CNT bundles grown by Thermal Chemical Vapor Deposition (TCVD) (**b**) Top view of MWCNTs-based device in CPW (coplanar waveguide) technology under a 125-μm-pitch test probe (optical microscope image).

For characterization technique improvements, a new differential-type approach is chosen by assuming a dedicated de-embedding structures definition. In order to overcome the physical constraints imposed by the material dimensions and the technological process, a transmission line circuit topology in CPW technology was selected, with a tapered-type profile which preserves electrical contact from standard coplanar access to the circular cross section of the MWCNT bundle.

3.2. Material Microwave Conductivity Extraction Procedure

From material considerations, insertion of the MWCNT bundle at the end of the electrode adds a complex impedance in series, which contributes to a resonant frequency shift as a reactance modification of the system. By a classical spatial integration of the bundle conductivity from (2) involving a bundle length L and diameter D, an equivalent impedance Z_{bundle} approach becomes a comprehensive representation of the added material at mesoscopic scale (4).

By introducing (2) in (4) and assuming that $Z_{bundle} = Z'_{bundle} + j\,Z''_{bundle}$ with a form factor $F = L/D$, real (Z'_{bundle}) and imaginary (Z''_{bundle}) parts of Z_{bundle} relations become inversely frequency dependent, as expressed in (5) and (6):

$$Z_{bundle} = \frac{1}{\sigma_{bundle}} \cdot \frac{4L}{\pi D^2} \quad (4)$$

$$Z'_{bundle} = \frac{\sigma_{b0}}{\sigma_{b0}^2 + \sigma_b(\omega)^2} \cdot \frac{4\pi F}{D} \quad (5)$$

$$Z''_{bundle} = -\frac{\sigma_b(\omega)}{\sigma_{b0}^2 + \sigma_b(\omega)^2} \cdot \frac{4\pi F}{D}. \tag{6}$$

4. Experimental Results

4.1. Technological Process

The vertical technology developed for device elaboration relies on a multistep process. In a first step, a 2-µm-thick layer of Mo was sputtered on a highly resistive (HR) Si/SiO_2 substrate for the definition of microwave structures. After optical lithography and ion beam etching (IBE), Al_2O_3 and Fe layers were locally deposited as catalyst for the thermal Chemical Vapor Deposition (TCVD) growth of MWCNT bundles at 700 °C, allowing for the growth of a bundle with a form factor F equal to 5.

4.2. Experimental Microwave Environment

A broadband microwave experimental setup based on a probe test equipment connected to a vectorial network analyzer was used for the on-wafer S-parameters measurements of devices after an on-wafer Short/Open/Load/Thru (SOLT) calibration on alumina substrate in the 0.2–67 GHz frequency band.

For microwave signal coupling methodology, different device inner electrodes (C1 to C5) were designed and tested in order to validate the material properties. Each design was replicated on the same wafer as two sets of samples were processed with and without CNT bundles on the same wafer, maintaining the same bundle diameter of 20 µm for each device, in order to extract the VA (vertically aligned) MWCNT bundle impedance contribution from the microwave test structure itself.

From reflection coefficient measurements, microwave input impedances Z_{in} and Z'_{in} were extracted from devices with and without VA MWCNT bundle implementation, as in Figure 2. As VA MWCNT bundles are electrically connected in series to microwave coplanar structures, the frequency-dependent complex impedance of VA MWCNT bundles was extracted from the reflection coefficient to input impedance conversion followed by a de-embedding technique assumed by direct impedance subtraction.

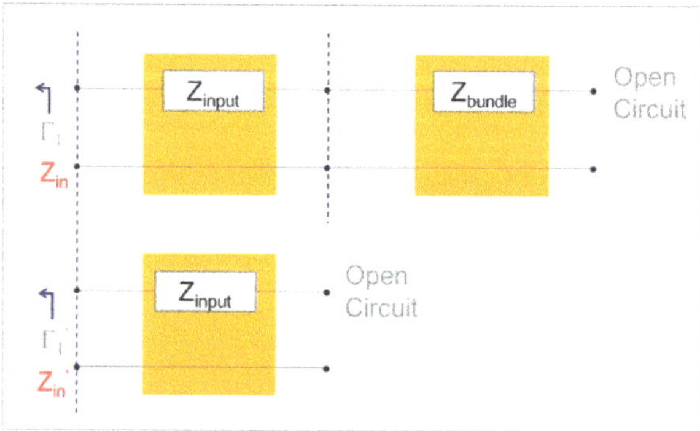

Figure 2. Schematic representation of the complex impedance extraction of VA (vertically aligned) MWCNT bundles from a two-device set.

As expected from theory and as shown in Figure 3a, a decrease of the real part of the impedance in frequency was confirmed from five samples measurements. At low frequencies, an impedance value of 80–100 Ω validates the collective response of the individual shunt high resistance of each MWCNT forming the bundle, with a density of 10^{15} units per cm^2. At higher frequencies, VA MWCNT bundles demonstrated a decreasing frequency-dependent impedance as expected from Equations (5) and (6), as well as a capacitive behavior from its negative imaginary part.

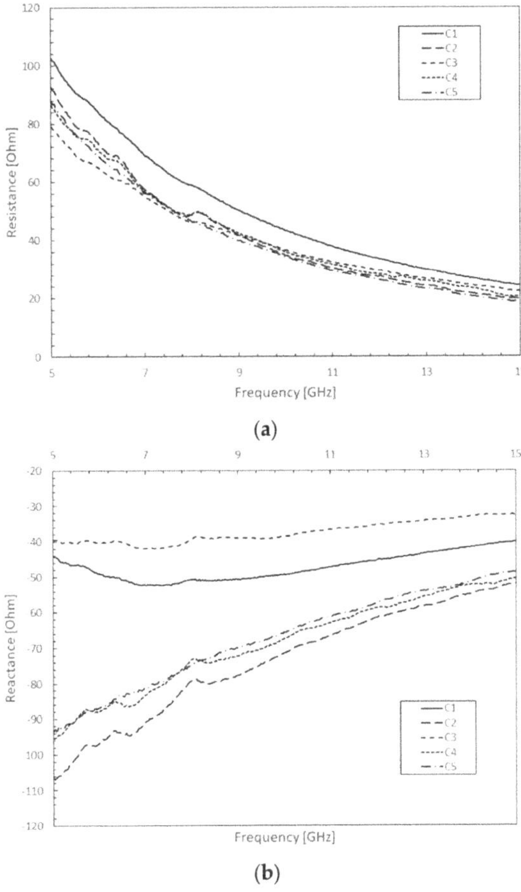

Figure 3. Experimental VA MWCNT bundle impedance extracted from five different CPW devices (C1 to C5) incorporating identical bundle dimensions in the 5–15 GHz frequency band: (**a**) Real part; (**b**) Imaginary part.

In addition, a non-negligible and complex conductivity in Figure 4 attributed to this material was experimentally observed from the five samples for the first time at microwave frequencies. These exceptional properties confirm this new material's implementation as a disruptive technology for the next generation of microwave devices designed with a high degree of miniaturization.

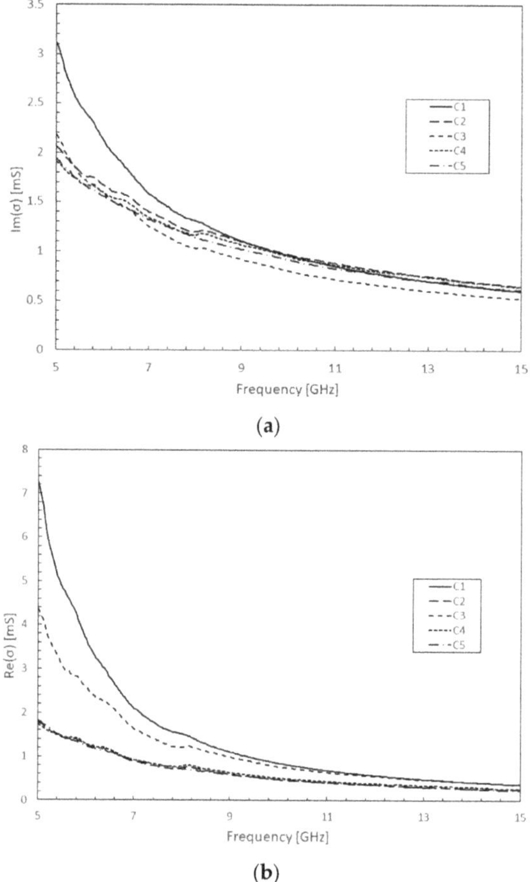

Figure 4. Experimentally extracted VA MWCNT complex conductivity measurements from five microwave devices (C1 to C5) incorporating identical bundle dimensions: (**a**) Real part; (**b**) Imaginary part.

5. Conclusions

Preliminary research works were performed on sub-wavelength MWCNT-based antennas designed benefiting from the natural inductive/capacitive behavior or complex conductivity never before achieved in classical conductors in the microwave domain. By also exploiting a vertically aligned CNT bundle configuration that drastically reduces the contact resistance of individual MWCNTs, the electromagnetic VA MWCNT bundle properties as obtained by equivalent complex impedance extraction from our experimental material process were identified for the first time using CPW technology, from 5 to 15 GHz frequency. These exceptional properties concretize the eligibility of this new material as a disruptive technology for the next generation of microwave devices and antennas, at the material level. Future works will focus on the determination of VA MWCNT bundle complex impedance law in respect of physical dimensions and cross section profile.

6. Patents

This work has led to the filing of a patent under deposition number FR1800496.

Author Contributions: Conceptualization, C.-T.C., J.C.; methodology, C.-T.C., S.X., Y.F.; investigation, C.-T.C., S.X., Y.F.; data curation, C.-T.C., J.-P.M.; writing, C.-T.C., J.C.; project administration, A.Z., J.C.; funding acquisition, A.Z., J.C.

Funding: This research was funded by the European Commission under grant FP7-ICT-2011-8 (project number 318352).

Conflicts of Interest: The authors declare no conflicts of interest. The funders had no role in the design of the study; in the collection, analyses, or interpretation of data; in the writing of the manuscript, or in the decision to publish the results.

References

1. Pesetski, A.A.; Baumgardner, J.E.; Folk, E.; Przybysz, J.X.; Adam, J.D.; Zhang, H. Carbon nanotube field-effect transistor operation at microwave frequencies. *Appl. Phys. Lett.* **2006**, *88*, 113103. [CrossRef]
2. Anantram, M.P.; Léonard, F. Physics of carbon nanotube electronic devices. *Rep. Prog. Phys.* **2006**, *69*, 507–561. [CrossRef]
3. Balasubramanian, K.; Burghard, M. Biosensors based on carbon nanotubes. *Anal. Bioanal. Chem.* **2006**, *385*, 452–468. [CrossRef] [PubMed]
4. Zhao, H.; Zhang, Y.; Bradford, P.D.; Zhou, Q.; Jia, Q.; Yuan, F.G.; Zhu, Y. Carbon nanotube yarn strain sensors. *Nanotechnology* **2010**, *21*, 305502. [CrossRef] [PubMed]
5. Schroeder, V.; Savagatrup, S.; He, M.; Lin, S.; Swager, T.M. Carbon nanotube chemical sensors. *Chem. Rev.* **2019**, *119*, 599–663. [CrossRef] [PubMed]
6. Farina, M.; Lucesoli, A.; Pietrangelo, T.; Di Donato, A.; Fabiani, S.; Venanzoni, G.; Mencarelli, D.; Rozzi, T.; Morini, A. Disentangling time in a near-field approach to scanning probe microscopy. *Nanoscale* **2011**, *3*, 3589–3593. [CrossRef] [PubMed]
7. Jin, X.; Hwang, J.C.M.; Mencarelli, D.; Pierantoni, L.; Farina, M. Nano Probing for Microwave ngineers: Calculating Probe-Sample Capacitance and Charge Distribution of a Near-Field Scanning Microwave Microscope on a Nanoscale. *IEEE Microw. Mag.* **2017**, *18*, 71–75. [CrossRef]
8. Dargent, T.; Haddadi, K.; Lasri, T.; Clément, N.; Ducatteau, D.; Legrand, B.; Tanbakuchi, H.; Theron, D. An interferometric scanning microwave microscope and calibration method for sub-fF microwave measurements. *Rev. Sci. Instrum.* **2013**, *84*, 123705. [CrossRef] [PubMed]
9. Legrand, B.; Salvetat, J.P.; Walter, B.; Faucher, M.; Théron, D.; Aimé, J.P. Multi-MHz micro-electro-mechanical sensors for atomic force microscopy. *Ultramicroscopy* **2017**, *175*, 46–57. [CrossRef] [PubMed]
10. Burke, P.J.; Li, S.; Yu, Z. Quantitative theory of nanowire and nanotube antenna performance. *IEEE Trans. Nanotechnol.* **2006**, *5*, 314–334. [CrossRef]
11. Hanson, G.W. Fundamental transmitting properties of carbon nanotube antennas. *IEEE Trans. Antennas Propag.* **2005**, *53*, 3426–3435. [CrossRef]
12. Tripon-Canseliet, C.; Xavier, S.; Modreanu, M.; Ziaei, A.; Chazelas, J. Vertically-grown MW CNT bundles microwave characterization for antenna applications. In Proceedings of the 2014 International Conference on Numerical Electromagnetic Modeling and Optimization for RF, Microwave, and Terahertz Applications (NEMO), Pavia, Italy, 14–16 May 2014; IEEE: Piscataway, NJ, USA, 2014.
13. Alu, A.; Engheta, N. Enhanced directivity from subwavelength infrared/optical nano-antennas loaded with plasmonic materials or metamaterials. *IEEE Trans. Antennas Propag.* **2007**, *55*, 3027–3039. [CrossRef]
14. Ziolkowski, R.W. Passive and active metamaterial-inspired nano-scale antennas. In Proceedings of the 10th European Conference on Antennas and Propagation (EuCAP), Davos, Switzerland, 10–15 April 2016.
15. Minatti, G.; Martini, E.; Maci, S. Efficiency of Metasurface Antennas. *IEEE Trans. Antennas Propag.* **2017**, *65*, 1532–1541. [CrossRef]
16. Mehdipour, A.; Rosca, I.D.; Sebak, A.R.; Trueman, C.W.; Hoa, S.V. Full composite fractal antenna using carbon nanotubes for multiband wireless applications. *IEEE Antennas Wirel. Propag. Lett.* **2010**, *9*, 891–894. [CrossRef]

Article

Fabrication of Microwave Devices Based on Magnetic Nanowires Using a Laser-Assisted Process

Vivien Van Kerckhoven [1], Luc Piraux [1] and Isabelle Huynen [2,*]

[1] IMCN Institute, Université catholique de Louvain, 1 place Croix du Sud, 1348 Louvain-la-Neuve, Belgium
[2] ICTEAM Institute, Université catholique de Louvain, 3 place du Levant, 1348 Louvain-la-Neuve, Belgium
* Correspondence: isabelle.huynen@uclouvain.be

Received: 11 June 2019; Accepted: 12 July 2019; Published: 16 July 2019

Abstract: This paper compares two laser-assisted processes developed by the authors for the fabrication of microwave devices based on nanowire arrays loaded inside porous alumina templates. Pros and cons of each process are discussed in terms of accuracy, reproducibility and ease of fabrication. A comparison with lithography technique is also provided. The efficiency of the laser-assisted process is demonstrated through the realization of substrate integrated waveguide (SIW) based devices. A Nanowired SIW line is firstly presented. It operates between 8.5 and 17 GHz, corresponding to the first and second cut-off frequency of the waveguide, respectively. Next, a Nanowired SIW isolator is demonstrated. It shows a nonreciprocal isolation of 12 dB (corresponding to 4.4 dB/cm), observed in absence of a DC magnetic field, and achieved through an adequate positioning of ferromagnetic nanowires inside the waveguide cavity.

Keywords: microwave; ferromagnetic; laser processing; substrate integrated waveguide; nanowire

---

1. Introduction

Porous templates loaded with nanowires have been extensively studied and exploited to fabricate planar and monolithic devices, thanks to the interaction between nanowires and microwave signals. Microwave devices based on nanowires have several advantages over classical components: wide range of used frequencies, stability over temperature, miniaturization through integration into a single substrate. Ferromagnetic nanowire arrays have higher saturation magnetization and ferromagnetic resonance (FMR) than classical ferrites. They can also operate at higher frequencies due to their large aspect ratio.

Owing to those properties, absorbers [1,2], inductors [3], noise suppressors [4] or filters based on electromagnetic band-gap effect [5], have been developed, as well as non-reciprocal devices, including circulators [6], isolators [7] and phase shifters [8]. More recently, the high permittivity of metallic nanowire arrays [9] was used to design slow-wave transmission lines [10], while the double ferromagnetic resonance effect in magnetic nanowires was exploited to design a filter [11]. Different nanowire materials have been combined with coplanar or microstrip topologies.

As outlined in Reference [12], high-quality transmission lines are necessary to match the constraints of low cost and consumption, high density and high operational frequency. Above 30 GHz, interferences and radiation losses deter the performances of classical microstrip and coplanar geometries. Hence, the substrate integrated waveguide (SIW) concept [13] is an interesting solution. Two rows of vertical metallic wires or metallized vias are drilled throughout the entire height of a substrate to obtain a rectangular section of shielded line, allowing to reduce radiation. The presence of wires also prevents spurious interferences by crosstalk due to electromagnetic coupling between adjacent devices in a same circuit. Crosstalk as low as −20 dB [14], −39 dB [15], or even −50 dB [16] are reported for SIW technology. Various active and passive devices can be integrated in the same substrate, including antennas, filters, couplers, power suppliers, for operation up to 180 GHz, and using different kinds of substrates (Printed Circuit Boards (PCB), paper, polymers or alumina) [17,18]. A review and recent developments on SIW antennas can be found in References [19–21]. For classical antennas, crosstalk is also a critical isssue since it is responsible for spurious coupling between adjacent patches in an array, resulting into a degradation of radiation performances. Solutions based on metamaterials [22,23] and periodic electromagnetic band gap (EBG) structures [24,25] have been proposed. These solutions will be briefly addressed for SIW antennas in Section 2.3.2.

The combination of nanowires and substrate integrated waveguide (SIW) devices is a promising approach, taking into account their respective advantages [12,26,27]. This concept uses metallic nanowire arrays deposited in a porous alumina template to form the waveguide walls. Then, Cu layers deposited on top and bottom surfaces of the alumina substrate form a simple rectangular waveguide. Various nanowire arrays can be added inside the waveguide cavity, combining different heights, shapes and materials (magnetic or not), to obtain competitive devices. Their realization requires to grow nanowires inside the template in specific positions with a good precision and accuracy.

In this paper we compare two laser-etching processes developed for the precise location of areas hosting the growth of nanowires in a porous alumina membrane. In the first method, the laser opens windows in a top metallic mask, in order to allow the growth in these areas. In the second method, the laser etches the surface of the pores in dedicated areas in order to destroy the porosity and prevent the growth of nanowires during the ECD (electrochemical deposition) process. The two methods are described in the following sections. Next, their application is illustrated with the fabrication of two Nanowired Substrate Integrated Waveguide devices (NSIWs).

2. Results and Discussion

2.1. Nanowired Template

A schematic of the porous template supporting the Nanowired Substrate Integrated Waveguide is shown in Figure 1. Some regions contain electrodeposited nanowire arrays loaded inside the alumina template. The procedure of electrodepositon is briefly described in Section 3.2. It allows a fine control of the height of deposited nanowires (NWs), while their diameter is fixed by the pore diameter of the membrane, as shown in Section 3.1.

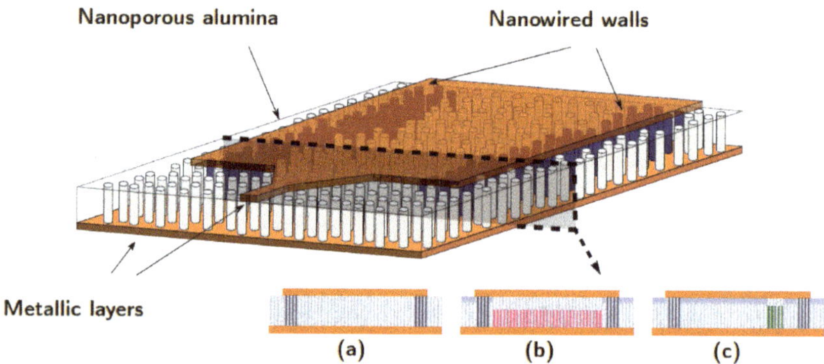

Figure 1. Schematic view of the Nanowire Substrate Integrated Waveguide (NSIW) in the alumina template, and different possible profile sections: (**a**) empty waveguide (**b**) homogeneous filling (**c**) asymmetric filling used for the isolator device. The proper placement of the nanowire arrays is achieved with the laser-assisted process described in Sections 2.3.1 or 2.3.2.

2.2. Design of NSIW Devices

The NSIW is constituted of two parallel rows of nanowire arrays separated by a distance a, embedded in a nanoporous template covered on both faces by 3 μm thick copper layer (see Figure 1). This structure constitutes a shielded rectangular cavity through which a microwave signal can propagate according to precise modes defined by the classical waveguide theory [28]. This signal only propagates above a cut-off frequency f_c depending on the waveguide width a and height b, on the medium relative permittivity ε_r and permeability μ_r and on the type of mode. For the first modes, f_c has the expression:

$$f_{c_{m0}} = c_0 \left(2ma \sqrt{\varepsilon_r \mu_r}\right)^{-1} \tag{1}$$

where c_0 is the speed of light in vacuum and m is the index of the mode. This cut-off frequency constitutes the lower bound of the single mode interval, where no other can create interferences. A frequency range exists where only one mode, referred as TE_{10} can appear. The latter has the lowest cut-off frequency.

The relative permittivity of nanoporous AAO is given by

$$\varepsilon_{r,AAO} = p\varepsilon_{r,air} + (1-p)\,\varepsilon_{r,Al_2O_3} \tag{2}$$

with p the porosity of the alumina template, $\varepsilon_{r,air} = 1$ the permittivity of the empty pores, $\varepsilon_{r,Al_2O_3} = \varepsilon_a\,(1 - j\tan\delta_a)$ the permittivity of bulk alumina, $\varepsilon_a = 9.8$ the dielectric constant of bulk alumina and $\tan\delta_a = 1.5 \times 10^{-2}$ its loss tangent factor.

For the dominant TE_{10} mode, the magnetic field inside the waveguide has the following expression:

$$H_x(x,y,z) = A\,\frac{jka}{\pi} sin \frac{\pi x}{a} \tag{3a}$$

$$H_y(x,y,z) = 0 \tag{3b}$$

$$H_z(x,y,z) = A\ cos \frac{\pi x}{a} \tag{3c}$$

where A is a constant and $k = 2\pi f \sqrt{\varepsilon_r \mu_r}/c_0$ is the propagation constant with f the operating frequency. The magnetic field is circularly polarized if

$$H_x(x,y,z) = \pm j H_z(x,y,z) \tag{4}$$

This occurs at a precise transversal position x_o, given by

$$x_0 = \pm \frac{a}{\pi} \arctan j \frac{\pi}{ka} \tag{5}$$

To obtain an isolator, a ferromagnetic material has to be inserted in the waveguide cavity at position x_0. Indeed, a circularly polarized magnetic field oscillating in a magnetized medium can induce ferromagnetic resonance: the tensorial permeability $\overline{\overline{\mu}}$ collapse in an effective scalar permeability noted μ_{eff}. This permeability seen by the circularly polarized wave is depending on the direction of propagation of the wave (along positive or negative longitudinal z-axis of SIW) and is written

$$\mu_{eff}^{\pm} = \mu \pm \kappa \tag{6}$$

where κ and μ hold for the off-diagonal and diagonal terms of the tensor $\overline{\overline{\mu}}$ describing the permeability of the ferromagnetic nanowire array placed in x_0. These components are given by [6]

$$\begin{aligned} \mu &= 1 + ph\frac{(\omega_r + j/\tau)\,\omega_M}{(\omega_r + j/\tau)^2 - \omega^2}\,\frac{(1+m^2)}{2} \\ \kappa &= ph\frac{\omega\omega_M}{(\omega_r + j/\tau)^2 - \omega^2}\,m \end{aligned} \quad (7)$$

where h stands for the nanowire array relative height, p for the nanowires packing factor which coincide with the porosity of the alumina template, $\omega_r = \omega_M(1-3p)/2$ for the resonance pulsation. $\omega_M = \gamma\mu_0 M_s$ is the Larmor pulsation, M_s is the saturation magnetization and $m = M/M_s$ the normalized remanence. Finally, the material properties γ and τ are the gyromagnetic ratio and the relaxation time, respectively [5,6], while $\omega = 2\pi f$.

Close to the resonance frequency, the circularly polarized wave, propagating towards the negative z-axis through the ferromagnetic slab placed inside the waveguide according to Equation (5), will experience a strong attenuation induced by the resonant imaginary part of μ^-_{eff}. On the other hand, the wave propagating towards the positive z-axis will be preserved since the imaginary part of μ^+_{eff} remains close to zero, inducing almost no attenuation. This is the desired non-reciprocal effect: the wave can easily travel the NSIW loaded with ferromagnetic nanowires in one direction, while being blocked in the other. The ratio between attenuations in the two directions is defined as isolation and is proportional to $\Im(\mu^-_{eff})$.

2.3. *Laser Etching Processes*

2.3.1. Laser Etching of Sacrificial Layer

This laser patterning process begins with the deposition of the cathode onto the bottom face of the template (Figure 2a). A 1.5 µm thick bottom metallic layer (Cr (5 nm)/Cu (1400 nm)/Au (100 nm)) is deposited by e-beam evaporation. It serves as cathode for the electrodeposition of nanowire arrays and as ground plane of the RF device. A 1 µm thick Al sacrificial layer is then obtained by sputtering onto the top of the AAO membrane in a separate step (Figure 2b). Then the first laser etching step (Figure 2c) is carried out on aluminium. The laser operation is described in Section 3.3. The following laser etching parameters are used: spacing between adjacent lines (2 µm), etching rate (200 mm·m^{-1}) and pulse repetition frequency (40 kHz). After the laser etching, nanowires are grown in the released pores by ECD described in Section 3.2 (Figure 2d,e). Various materials can be deposited in different released areas and different heights can be reached. The ECD of the nanowires is not affected by the aluminium layer on top of the membrane unless they overflow the nanopores. In such a case, a short circuit appears between the cathode and the aluminium layer, rendering electrodeposition in the pores much less effective. Hence, in the case of the pattern used in the process described by the Figure 2, the steps c and b could not have been inverted. If the overfilled nanowires are grown first, it will not be possible to make the second nanowire arrays. Afterwards the aluminium layer is totally removed (Figure 2f). It is worth mentioning that the laser beam cannot pass on the filled sections as it affects the nanowires contained in the template and slightly modify their magnetic properties. The result is an AAO membrane loaded with two nanowires arrays constituted of different materials and having different thicknesses. Finally, a second deposition of metallic layer is performed on the top side of the filled membrane, in order to realise the RF devices via a laser-etching of adequate metallic patterns for RF devices, as will be illustrated in Figures 4a and 5b. For convenience, the method is noted LES for laser etching of sacrificial layer.

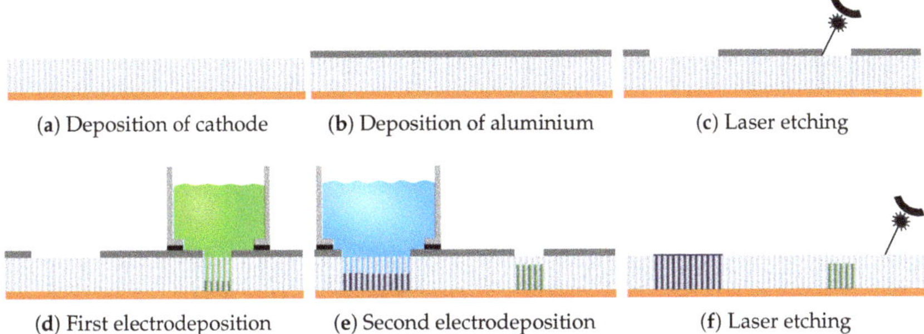

Figure 2. Schematic description of laser patterning process based on the etching of a sacrificial layer (LES) applied to a device containing two materials of different heights. (**a**) Deposition of the copper cathode at the bottom of the alumina membrane. (**b**) Deposition of the 1000 nm thick aluminium layer on the top face of the alumina membrane. (**c**) the laser both defines and etches the desired pattern. (**d**) electrodeposition is used for filling alumina template inside porous regions. The growth area is defined by the etched aluminium while the black rubber is used as a seal. Here, the electrochemical deposition (ECD) is carried out until the nanowires (NWs) reach half of the membrane height. (**e**) Electrodeposition is performed in another area until complete filling of pores and metal overflow. In this scheme, the nanowires are connecting the two metal electrodes. (**f**) Complete removal of the aluminium layer on top face of unfilled porous regions.

The LES process suffers from a lack of stability: uncontrolled small variations of the laser power generate dramatic changes in the efficiency of the aluminium etching process. Reductions induce a risk of an underetching of the sacrificial layer preventing the ECD. On the other hand, the template surface porosity can be destroyed by a too large laser power, preventing again the ECD. Section 2.3.2 proposes a simple and flexible laser-assisted procedure, based on the localized controlled destruction of the surface porosity. The electrolyte only enters in non-destroyed porous areas and small variations of laser power have no effect on their definition.

2.3.2. Laser Etching of Pores

In order to overcome the issues presented above with the LES process, we propose here a second procedure enabling the controlled destruction of the porous template surface over well-defined regions. In this scheme, the electrolyte only enters in non-damaged porous areas thus providing a simple and flexible method for a spatial control of NW growth. This process is noted LEP for laser etching of pores. Moreover, the ECD process is not affected by the small variations of laser power. Also, no additional etching step has to be performed, thus keeping the NWs far from laser beam. Finally, it also allows complex patterns to be generated for various microwave devices.

As in the previous procedure, the first experimental step is the deposition of the bottom metallic layer, which will act as a cathode for nanowires growth and as a ground plane for RF measurements (Figure 3a). Then, we carry out a laser etching of porous areas on the top face of the membrane (Figure 3b). In order to damage sufficiently the AAO template surface without removing a substantial thickness, we make three passes at a power of 0.10 W, a pulse frequency of 40 kHz, with a speed of 200 mm/s, and a distance between the lines of 5 µm. These parameters allow a homogeneous distribution of the laser spots in the two directions of the etched layer. After the etching, a small thickness of alumina has been removed in the illuminated parts. These parameters are a tradeoff between several constraints. First, the etching must not sublimate the alumina over a too large thickness. This could be a problem for the RF device design. Secondly, the power must remain relatively small enough to not degrade the cathode through the alumina. A damaged cathode could also cause RF losses. These first two conditions require the use of a relatively low laser power. Hence, passing several times is necessary to ensure the clogging of the pores. Finally, the areas to be engraved are larger than in the case of the first method. It is therefore necessary to have a high engraving speed and a relatively low path density so that engraving does not take too long.

After laser etching, the nanowires growth by ECD takes only place in the pores that have not been damaged by laser beam. It allows then a precise control of their positioning. Different materials can be deposited inside the same template using several ECD steps. The height of the nanowires can also be controlled by the deposition time to meet the needs. In contrast to the former LES method, the ECD in the LEP process may be continued after overfilling of the pores. As a result, the order in which the ECD steps are conducted to create the example pattern reported in Figure 3 can be inverted. Finally, as in Section 2.3.1, RF patterns are laser-etched in the top metallic layer.

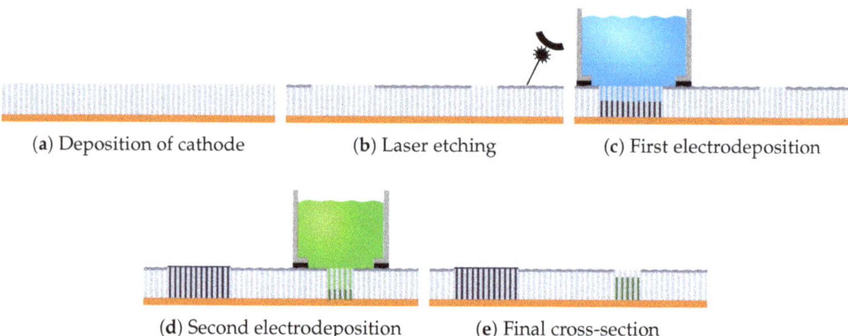

Figure 3. Laser process based on the Etching of the template surface Porosity (LEP), for a device including two materials of different heights. (**a**) E-beam deposition of Copper cathode. (**b**) Localised laser destruction of porosity on the top face. (**c**) ECD of first material until overflowing. (**d**) ECD of second material until a partial height in unclogged pores. Unlike the LES method, the order of the ECD can be inverted. (**e**) Final cross-section of the AAO membrane.

In Figure 4, several images of results obtained with the LEP process are presented. In Figure 4a, an example of interdigitated electrodes (IDE) pattern after etching but before ECD is given. The etched part is darker while the bright color indicates the un-attacked pores ready for ECD. This shows an example of complex pattern of grown nanowires that can be created. Figure 4b,c are two optical microscope (OM) images of cross section of a 70% filled and a 100% filled membranes. We see the sharp transition between filled (black) and unfilled (blue) regions and the uniform filling along the whole cross-section. A CCD photograph of a microwave device (substrate integrated waveguide (SIW) with a EBG filter effect) is given Figure 4d. The darker areas are filled with nanowires. The width between the walls of the waveguide (i.e., the two long nanowired rows) is 6 mm. Six rectangular nanowired areas are posted in the guide in order to create an Electromagnetic Band Gap effect similar to Reference [5]. Finally, the two SEM images in Figure 4e,f give a view of the surface after the etching: the surface porosity is completely destroyed.

As briefly mentioned in the introduction, EBG structures are relevant in the filters and antennas fields. The same statement applies for the SIW technology. In References [29,30], a 2-D periodic EBG structure was implemented around an array of 4 SIW antennas to reduce spurious coupling and enhance the performances. Topologies of EBG SIW filters [31,32] or delay lines [33] were also proposed. SIW EBG filters and antennas could be easily implemented using nanowires and the LES or LEP method presented in this paper.

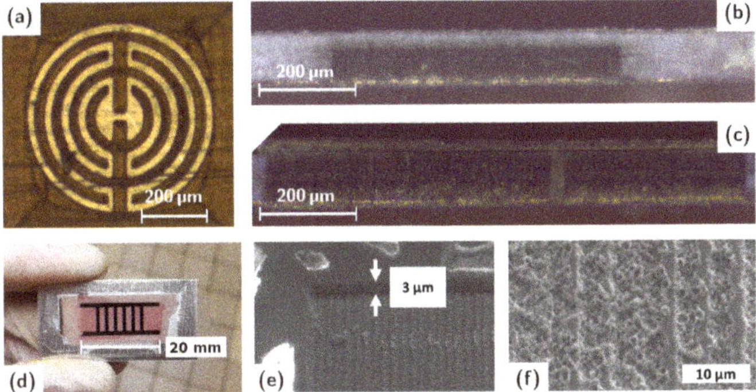

Figure 4. Devices realized with LEP process, at different steps. (**a**) Optical microscope (OM) image of IDEs after etching and before ECD. (**b**) OM image of an AAO template filled with NiFe nanowires at 70% of thickness of the template. (**c**) OM image of a template filled with Cu nanowires at 100%. (**d**) CCD camera image of a SIW device with EBG effect after ECD. Black areas are the nanowire-filled sections. The width between the two long lines forming the walls is 6 mm. (**e**) Scanning electron microscope (SEM) image at the edge of the etched area. The sublimated thickness is about 3 μm. (**f**) SEM image of the rough surface of the etched area.

2.3.3. Comparison between LES and LEP Processes

The LES process presented in Section 2.3.1 suffers from several issues. First, its feasibility is sensitive to the stability of the laser source power. Indeed the growing areas are determined by the etching of the sacrificial layer. If the power is too low, the metal layer will be underetched, and remains a barrier against the ECD. Conversely, if the power is too high, the porosity of the membrane is destroyed, preventing again the ECD. In the two cases a fine tuning and readjustement of the laser power is necesssay, affecting the good reproducibility of the process. On the contrary in the LEP method presented in Section 2.3.2 the laser power does not affect the ECD, since it aims to destroy only the pore areas where ECD must be prevented.

The second drawback of the LES method of Section 2.3.1 is related to the thickness and nature of the sacrificial layer necessary to prevent ECD in the areas where NWs are not needed. We used an aluminum thickness of 1000 nm to completely seal the pores of the templates in areas where ECD must be avoided. Etching such a thickness requires to use a high number of passes of the beam over a same track since the laser power cannot be increased too much as explained above. Hence, the sources of errors are multiplied and the irreproducibility is enhanced.This drawback is not present in the second method LEP since there is no need for etching of a sacrificial layer before ECD.

Another advantage of the LEP method in Section 2.3.2 is that the order of the various ECDs is not important. The ECD of overfilled nanowires in some areas of the template—which is mandatory for the realisation of NSIW devices illustrated in Figure 1—does not prevent next ECDs of nanowires having smaller height.

Finally, as the initial sacrifial layer is not present in the LEP method, there is no risk to damage the magnetic properties since the final etching step of Figure 2f is absent. Our studies have indeed shown that the laser beam illuminating a nanowired area dramatically affects the apparent saturation magnetisation of the nanowires, lowering its value by at least 20%. This is due to a significant loss of magnetic material during laser ablation, which may be critical for the operation of microwave nonreciprocal devices.

For the sake of completeness Table 1 provides an additional comparison between laser process and classical lithography technique.

Table 1. Comparison between lithography and laser processes.

Technology	Lithography	Laser
accuracy	good (<100 nm)	good (<1000 nm)
reproducibility	good (mastered process)	poor for LES (unstability of power source)
cost	high (fabrication of masks)	cheaper
handling	intricate	easy and fast
versatility	poor (1 mask needed per set of devices)	easy reconfigurable for each size and topology
risks	clogging of pores by resins	impact of rugosity on RF losses

2.4. *Microwave Devices*

2.4.1. NSIW Line

The geometry of the NSIW line is depicted in Figure 1. The nanoporous alumina template described in Section 3.1 is filled with ferromagnetic nanowires (NWs) according to the ECD process in Section 3.2. Figure 1b shows the cross-section of the NSIW homogeneously filled with NWs, while Figure 1c corresponds to an NSIW filled in selective areas using the LEP process described in Section 2.3.2.

The measured transmission S_{21} through the empty NSIW (Figure 1a) is shown in Figure 5. The measurement method is described in Section 3.4. Shielding walls are made with Cu nanowires, the waveguide cavity size is 6×12 mm^2. The cut-offs of the 3 first modes TE_{10}, TE_{20} and TE_{30} appear at the values predicted by the theory (1): 8.5 GHz, 17 GHz and 25 GHz.

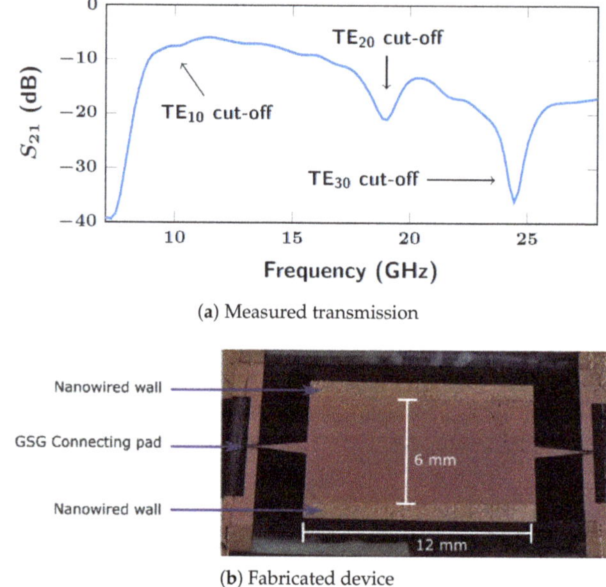

(**a**) Measured transmission

(**b**) Fabricated device

Figure 5. (**a**) Measured transmission through a 12 mm long NSIW. Two walls made of Cu nanowires define a 6 mm wide cavity guiding the signal and inducing a TE_{10} cut-off effect at 8.5 GHz, as predicted by the theory (1). (**b**) photograph of the fabricated device, showing the NSIW cavity of length 12 mm embedded between two triangular tapered microstrip transitions, each connected to a rectangular contact pad for the coplanar waveguide (CPW) measuring probe.

2.4.2. NSIW Isolator

An NSIW isolator was also designed and fabricated using the LEP process in Section 2.3.2, and measured. The cavity is 6 mm-wide and 20 mm-long. A 10 mm-long and 1.5 mm-large area of ferromagnetic NiFe nanowire arrays was grown inside the cavity (see Figure 1c) at the transversal position given by (5), x_0 = 1.6 mm. The relative height of nanowires is 70%. We used saturation magnetization M_s = 1000 kA·m^{-1}, remanence factor m = 0.8, the gyromagnetic factor $\gamma = 3.08 \times 10^{10}$ s·A·kg^{-1} and relaxation time τ = 7 × 10^{11} s. The ferromagnetic nanowired-filled area inside the waveguide is 10 × 1.5 mm^2 = 15 mm^2. The difference between the forward S_{21} and backward S_{12} transmissions, i.e., the isolation, is shown on Figure 6. A maximum occurs at 13 GHz, which is induced by the ferromagnetic resonance of NiFe material, and is the signature of a nonreciprocal propagation through the device, since S_{21} and S_{12} are significantly different (12 dB).

The LEP process preserves the porosity in areas where ECD occurs, hence the amount of nanowires in the template is uniform over the whole area of the membrane, other said the targed filling factor of 70% is reached uniformly in the template. As a consequence, the value of permeability $\Im(\mu^-_{eff})$ (6) remains well controlled, as well as finally the level of isolation.

The insertion losses observed in Figure 5 (8 dB at 12 GHz, corresponding to 4.4 dB/cm) are comparable or lower than other technologies/topologies, as shown Table 2. They should be significantly decreased by further improvement of the devices. Indeed, the upper and lower Cu layers deposited on the two faces of the AAO membrane can be thickened to increase the shielding effect. Also, the amount of magnetic material placed inside the isolator can be increased by enlarging the width of the NiFe NW strip. All those improvements will enhance significantly the performances of NSIW devices in terms of insertion losses in order to achieve insertion losses lower than 1 dB/cm.

Table 2. Comparison of insertion losses (IL).

Ref.	Topology	Technology	f (GHz)	IL (dB/cm)
[34]	CPW	Ni_xSi_y/Si	10	2
[35]	CPW	MMIC	20	5
[36]	SIW	LTCC	50	2.53
[37]	SIW	LTCC	38	4
[10]	microstrip	NW in AAO	60	7
[38]	SIW	NW in AAO	20	7.4
present work	SIW	NW in AAO	12	4.4

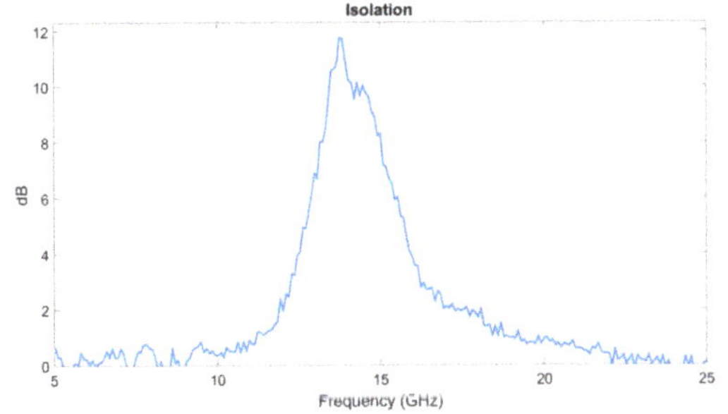

Figure 6. Difference between the forward S_{21} and backward S_{12} transmissions, i.e., isolation, obtained through a NSIW isolator. The non-reciprocal effect is created by an area of NiFe nanowires asymmetrically placed inside the waveguide cavity (see Figure 1c). The NSIW isolator is fabricated using the LEP process with 1.5 mm-wide NiFe area.

3. Materials and Methods

3.1. Nanoporous Template

The porous templates used for this work are anodized aluminium oxide membranes (AAO) from Smart Membranes GmbH (Figure 7). The pore diameter is 40 nm, the distance between pores is 125 nm and the thickness is 100 μm.

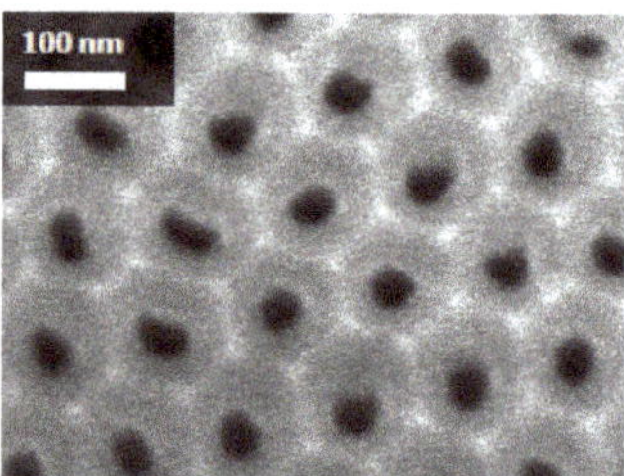

Figure 7. SEM images of a self-supported AAO templates with pore diameter of 40 nm, interpore distance of 125 nm, and thickness of 100 μm.

3.2. Electrochemical Deposition Process (ECD)

Prior to the ECD process, a thick metallic layer is deposited on one face of the AAO template, to completely cover the nanopores and play the role of cathode during the electrodeposition. The usual cathode is a three-layer (Cr (5 nm)/Cu (1400 nm)/Au (100 nm)) deposited by e-beam evaporator. A schematic representation of the experimental set-up is shown in Figure 8. The alumina membrane is placed in a Teflon cell that contains the electrolytic solution. We used an AgCl reference electrode and a Pt wire as anode The three-electrodes are connected to a EG & G Princeton Applied Research potentiostat.

Different materials were deposited inside the same porous template using successive electroplating steps. The height of the nanowires was controlled by the deposition time (see Figure 4), modifying subsequently the permittivity ε of the filled template sections [9]. In this work, we used two different electrolytic solutions:

- for NiFe nanowires:
  (1 M $NiSO_4$ + 0.02 M $FeSO_4$ + 0.5 M H_3BO_3), V = −1.05 V
- for Cu nanowires:
  (1 M Cu basis + 0.08 M H_2SO_4 + chloride basis + organic additives), V = −0.015 V
  The detailed composition of the Cu solution is a property of Sigma-Aldrich, Inc.

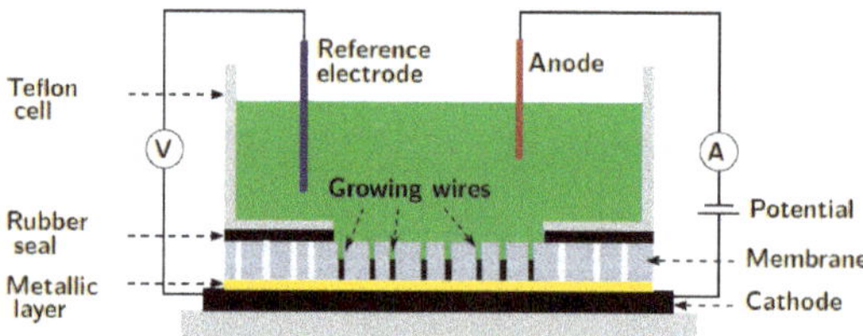

Figure 8. Schematic illustration of the electrodeposition cell using the three-electrodes configuration (i.e., three-layered cathode, Pt anode and AgCl reference). The membrane is covered by the Teflon cell containing the electrolyte, over an area where the pores are opened. A piece of rubber is inserted for ensuring the sealing. Adapted from Reference [39].

3.3. Pico-Second Laser System

The used laser is a Q-Switch picosecond laser assembled by Oxford Lasers Ltd. [40], powered with laser source fabricated by Coherent, Inc, operating in the ultraviolet (UV) range at 355 nm wavelength. Its use was previously illustrated for the micromachining of carbonated nanostructures [41]. This "pico-second" corresponds to the order of magnitude of the duration of the laser pulses and is therefore not directly related to their frequency. Due to the UV light absorption difference between the aluminium and alumina, alumina is less damaged with the etching. The pulses repetition frequency is basically fixed at 200 kHz and the 100% laser power is close to 4 W. These parameters can be lowered to meet the needs. To fully remove aluminium, a power of 0.05 W and 3 passes are necessary and these parameters prevent cathode damage. The laser etches clean and sharp edges of metal. The resulting aluminium walls show micron size surface roughness and the pores are totally opened in the etched areas [27], so that the electrolyte can easily penetrate into the AAO template for the subsequent EDC.

The selection of the laser parameters used to remove a 1000 nm thick layer of aluminium deposited on a nanonanoporous alumina membrane without damaging it is as follows. First, for simplicity, vertical defocusing is set to zero because its effects are difficult to predict. We aim to have the smoothest and homogeneous possible etching. This is possible by imposing a uniform distribution of laser spots in the horizontal plane, in both X and Y directions. The distance between the lines—denoted e—must therefore correspond to the distance between two spots on the same line. This implies the relation:

$$e = \frac{v}{f_r} \tag{8}$$

where v is the speed of the spot and f_r is the pulse repetition frequency. The distance e is set first. Considering that the diameter of the laser spot is 10 μm and that an overlap is necessary to ensure a correct ablation of the aluminium, we start by imposing e = 5 μm. Several tests performed with this value have shown that, for a thickness of 1000 nm, the induced overlap is not sufficient (see Figure 9a). Indeed, as the passes go by, the lines on which the spot center pass are more intensely engraved than the areas between these lines. This result is due to two effects: the inaccuracy of the laser spot longitudinal position on the one hand, and the non-homogeneity of the laser power on the spot surface on the other hand. The power in the laser beam follows a Gaussian distribution. In order to increase the recovery and reduce the impact of these two factors, the value of 2 μm was chosen for e. This provides better homogeneity in the etching (see Figures 9b and 10). Experience shows that if the laser spot moves too fast, the definition of the followed path is not accurate, especially around corners and half-turns. This definition comes acceptable when the speed approaches 300 mm/s. We chose 200 mm/s, which is a commonly used value. Hence, according to Equation (8), the frequency f_r must be 100 kHz.

The power and number of passes still have to be determined. Different experiments have led to the conclusion that passing three times (n = 3) over the aluminium layer, with a power corresponding to 4% of the internal power (which corresponds to 0.05 W if f_r is 100 kHz) provides very good results. Using lower power does not effectively sublimate the aluminium while reducing the number of passes doesn't give the possibility to finely control the engraved depth. Figure 9 was obtained on a square engraved with these parameters.

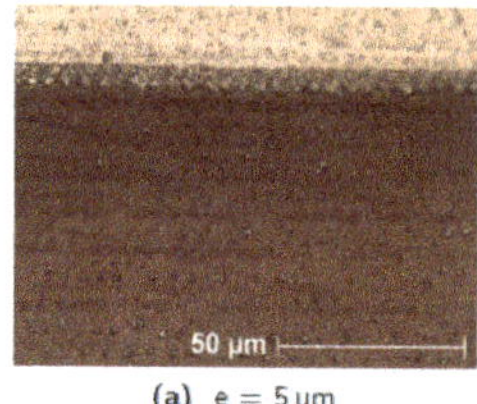

(a) e = 5 µm

(b) e = 2 µm

Figure 9. Optical images after etching of an Al 1000 nm layer taken with e = 5 µm (**a**) and e = 2 µm (**b**). The white material is aluminium and the brown color comes from the cathode seen through the AAO. A greater distance e allows horizontal lines to appear, parallel to the engraving paths, separated from 5 µm, indicating insufficient overlap. If e is reduced, recovery is better and homogeneity is significantly improved. The definition of the edge of the cleaned area is also much better with e = 2 µm. The etching parameters are f_r = 100 kHz, P = 0.05 W, n = 3.

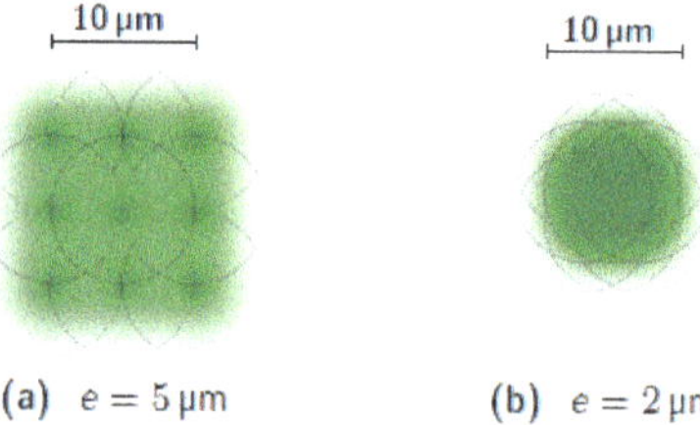

(a) $e = 5\,\mu m$ (b) $e = 2\,\mu m$

Figure 10. Schematic distribution of laser power on two illuminated areas with two values of e, the distance between spots. The intensity in the spot respects a Gaussian distribution. By decreasing e, the overlap and the homogeneity of the etching are increased but the etched surface is reduced.

3.4. Microwave Characterization

The standard microwave measurements were conducted with a microwave prober PM8 built by FormFactor, Inc. The microwave transmission is measured with a vector network analyser (VNA) model N5247A PNA-X from Agilent Technologies. The devices under test (DUTs) were connected thanks to coplanar GSG probes that had a 150 µm pitch, fixed on the prober and connected to the VNA via high performances semi-rigid RF cables. GSG probes are used to excite a tapered microstrip line feeding the NSIW devices. The coplanar waveguide (CPW) ports and the microstrip lines are designed to be matched to 50 Ω, while the dimensions of the tapered transitions from microstrip to SIW have been optimized numerically to prevent the generation of spurious propagation modes. The VNA measures the S-parameters of the devices at 401 frequency points homogeneously ranged from 100 MHz up to 40 GHz. Prior to measurements, the setup was calibrated with a SOLT method [42] using a standard planar calibration kit. The de-embedding is executed by the VNA itself. This way, the reference planes between which the transmission is determined are moved at the end of the two RF probes.

4. Conclusions

Various structures based on nanoporous templates have been studied in the past. In particular several devices were developed using templates filled with ferromagnetic nanowires for operation in the microwave frequency range. This paper compares two laser-assisted processes developed by the authors for the fabrication of various microwave devices exploiting nanoporous alumina templates filled with ferromagnetic nanowires. The pros and cons of each process are discussed. The efficiency of the laser-assisted process is demonstrated through the realization of various devices, showing performances in line with the literature. They combine ferromagnetic nanowires with the Substrate

Integrated Waveguide topology. The present work paves the road towards the realization of tunable SIW antennas and filters, by exploing the tunability of the FMR phenomenon via an external DC magnetic field, as we already demonstrated for the microstrip technology [11].

Author Contributions: Design, fabrication and experimental characterization, writing: V.V.K. Writing and supervision: L.P. and I.H.

Funding: This work was supported by the Research Science Foundation of Belgium (FRS-FNRS). Vivien Van Kerckhoven acknowledges the FRS-FNRS for financial support (FRIA grant). Isabelle Huynen is Research Director of FRS-FNRS.

Acknowledgments: The authors would like to thank Ester Tooten, Pascal Simon and Pascal Van Velthem for their help and advices.

Conflicts of Interest: The authors declare no conflict of interest.

Abbreviations

The following abbreviations are used in this manuscript:

AAO	Anodic Aluminium Oxide
CCD	Charge-Coupled-Device
CPW	Coplanar Waveguide
Cu	Copper
DUT	Device Under Test
EBG	Electromagnetic Band Gap
ECD	Electro-Chemical Deposition
GSG	Ground-Signal-Ground
IDE	Interdigitated Electrode
NiFe	nickel-iron alloy
MMIC	Monolithic Microwave Integrated Circuit
LEP	Laser Etching of Pores
LES	Laser Etching of Sacrificial layer
LTTC	Low Temperature Co-fired Ceramics
NW	Nanowire
NSIW	Nanowired Substrate Integrated Waveguide
OM	Optical Microscope
PCB	Printed Circuits Boards
RF	Radio Frequency
SEM	Scanning Electron Microscope
SOLT	Short-Open-Line-Thru
SIW	Substrate Integrated Waveguide
VNA	Vector Network Analyzer

References

1. Qiao, L.; Han, X.; Gao, B.; Wang, J.; Wen, F.; Li, F. Microwave absorption properties of the hierarchically branched Ni nanowire composites. *J. Appl. Phys.* **2009**, *105*, 053911. [CrossRef]
2. Chen, W.; Han, M.; Deng, L. High frequency microwave absorbing properties of cobalt nanowires with transverse magnetocrystalline anisotropy. *Phys. B Condens. Matter* **2010**, *405*, 1484–1488. [CrossRef]
3. Hamoir, G.; Piraux, L.; Huynen, I. Q-factor improvement of integrated inductors using high aspect ratio ferromagnetic nanowires. *Microw. Opt. Technol. Lett.* **2012**, *54*, 1484–1488. [CrossRef]
4. Kuanr, B.; Veerakumar, V.; Marson, R.; Mishra, S.; Kuanr, A.; Camley, R.; Celinski, Z. Nickel Nano-Wires Filled Alumina Templates for Microwave Electronics. *IEEE Trans. Magn. Lett.* **2009**, *45*, 4052–4055. [CrossRef]
5. Saib, A.; Vanhoenacker-Janvier, D.; Huynen, I.; Encinas, A.; Piraux, L.; Ferain, E.; Legras, R. Magnetic photonic band-gap material at microwave frequencies based on ferromagnetic nanowires. *Appl. Phys. Lett.* **2003**, *83*, 2378. [CrossRef]

6. Saib, A.; Darques, M.; Piraux, L.; Vanhoenacker-Janvier, D.; Huynen, I. An unbiased integrated microstrip circulator based on magnetic nanowired substrate. *IEEE Trans. Microw. Theory Tech.* **2003**, *83*, 2378. [CrossRef]
7. Carignan, L.; Caloz, C.; Menard, D. Dual-band integrated self-biased edge-mode isolator based on the double ferromagnetic resonance of a bistable nanowire substrate. In Proceedings of the 2010 IEEE MTT-S International Microwave Symposium, Anaheim, CA, USA, 23–28 May 2010; pp. 1336–1339.
8. Hamoir, G.; Medina, J.D.; Piraux, L.; Huynen, I. Self-Biased Nonreciprocal Microstrip Phase Shifter on Magnetic Nanowired Substrate Suitable for Gyrator Applications. *IEEE Trans. Microw. Theory Tech.* **2012**, *60*, 152–2157. [CrossRef]
9. Spiegel, J.; de la Torre, J.; Darques, M.; Piraux, L.; Huynen, I. Permittivity Model for Ferromagnetic Nanowired Substrates. *IEEE Microw. Wirel. Compon. Lett.* **2007**, *60*, 492–494. [CrossRef]
10. Serrano, A.; Franc, A.-L.; Assis, D.; Podevin, F.; Rehder, G.; Corrao, N.; Ferrari, P. Slow-wave microstrip line on nanowire-based alumina membrane. In Proceedings of the 2014 IEEE MTT-S International Microwave Symposium (IMS2014), Tampa, FL, USA, 1–6 June 2014; pp. 1–4.
11. Ramirez-Villegas, R.; Huynen, I.; Piraux, L.; Encinas, A.; Medina, J.D.L.T. Configurable Microwave Filter for Signal Processing Based on Arrays of Bistable Magnetic Nanowires. *IEEE Trans. Microw. Theory Tech.* **2016**, *65*, 72–77. [CrossRef]
12. Van Kerckhoven, V.; Piraux, L.; Huynen, I. A novel laser-assisted fabrication process for nanowired substrate integrated devices. In Proceedings of the 2018 13th European Microwave Integrated Circuits Conference (EuMIC), Madrid, Spain, 23–25 September 2018; pp. 170–173.
13. Wu, K.; Deslandes, D.; Cassivi, Y. The substrate integrated circuits—A new concept for high-frequency electronics and optoelectronics. In Proceedings of the 6th International Conference on Telecommunications in Modern Satellite, Cable and Broadcasting Service 2003—TELSIKS 2003, Nis, Yugoslavia, 1–3 October 2003.
14. Pazian, M.; Bozzi, M.; Perregrini, L. Crosstalk in Substrate Integrated Waveguides. *IEEE Trans. Electromagn. Compat.* **2015**, *57*, 80–86. [CrossRef]
15. Simpson, J.J.; Taflove, A.; Mix, J.A.; Heck, H. Substrate integrated waveguides optimized for ultrahighspeed digital interconnects. *IEEE Trans. Microw. Theory Tech.* **2006**, *54*, 1983–1990. [CrossRef]
16. Lai, Q.H.; Fumeaux, C.; Hong, W. Mutual coupling between parallel half-mode substrate integrated waveguides. In Proceedings of the 2012 International Conference on Microwave and Millimeter Wave Technology (ICMMT), Shenzhen, China, 5–8 May 2012; Volume 5, 4p.
17. Bozzi, M.; Georgiadis, A.; Wu, K. Review of substrate-integrated waveguide circuits and antennas. *IET Microw. Antennas Propag.* **2011**, *5*, 909. [CrossRef]
18. Moro, R.; Bozzi, M.; Collado, A.; Georgiadis, A.; Via, S. Plastic-based Substrate Integrated Waveguide (SIW) components and antennas. In Proceedings of the 2012 7th European Microwave Integrated Circuit Conference, Amsterdam, The Netherlands, 29–30 October 2012; pp. 627–630.
19. Djerafi, T.; Doghri, A.; Wu, K. *Handbook of Antenna Technologies*; Springer Science+Business Media: Singapore, 2005; Chapter 5.
20. Debnath, P.; Chatterjee, S. Substrate integrated waveguide antennas and arrays. In Proceedings of the 2017 1st International Conference on Electronics, Materials Engineering and Nano-Technology (IEMENTech) (ICMMT), Kolkata, India, 28–29 April 2017; 6p.
21. Kumar, A.; Raghavan, S. A Review: Substrate Integrated Waveguide Antennas and Arrays. *J. Telecommun. Electron. Comput. Eng.* **2016**, *8*, 95–104.
22. Alibakhshikenari, M.; Virdee, B.S.; Shukla, P.; See, C.H.; Abd-Alhameed, R.; Khalily, M.; Limiti, E. Interaction Between Closely Packed Array Antenna Elements for Applications Such as MIMO Systems and Synthetic Aperture Radars. *Radio Sci.* **2018**, *53*, 1368–1381. [CrossRef]
23. Alibakhshikenari, M. Beam-scanning leaky-wave antenna based on CRLH-metamaterial for millimetre-wave applications. *IET Microw. Antennas Propag.* **2018**, *12*, 2241–2447. [CrossRef]
24. Alibakhshikenari, M.; Virdee, B.S.; See, C.H.; Abd-Alhameed, R.A.; Falcone, F.; Limiti, E. Mutual-Coupling Isolation Using Embedded Metamaterial EM Bandgap Decoupling Slab for Densely Packed Array Antennas. *IEEE Access* **2019**, *7*, 5182–51840. [CrossRef]
25. Alibakhshikenari, M.; Khalily, M.; Virdee, B.S.; See, C.H.; Abd-Alhameed, R.A.; Limiti, E. Mutual coupling suppression between two closely placed microstrip patches using EM-bandgap metamaterial fractal loading. *IEEE Access* **2019**, *7*, 23606–23614. [CrossRef]

26. Van Kerckhoven, V.; Piraux, L.; Huynen, I. Substrate integrated waveguide isolator based on ferromagnetic nanowires in porous alumina template. *Appl. Phys. Lett.* **2014**, *105*, 183107. [CrossRef]
27. Van Kerckhoven, V.; Piraux, L.; Huynen, I.; A laser-assisted process to produce patterned growth of vertically aligned nanowire arrays for monolithic microwave integrated devices. *Nanotechnology* **2016**, *65*, 72–77. [CrossRef]
28. Pozar, D. *Microwave Engineering*, 4th ed.; Wiley: Hoboken, NJ, USA, 2012 .
29. Ashraf, N.; Haraz, O.; Ashraf, M.A.; Alshebeili, S. 28/38-GHz dual-band millimeter wave SIW array antenna with EBG structures for 5G applications. In Proceedings of the International Conference on Information and Communication Technology Research (ICTRC), Abu Dhabi, UAE, 17–19 May 2015; 4p.
30. Ghahramani, E.; Sadeghzadeh, R.A.; Boroomandisorkhabi, B.; Karami, M. Reducing mutual coupling of SIW slot array antenna using uniplanar compact EBG (UC-EBG) structure. In Proceedings of the 8th European Conference on Antennas and Propagation (EuCAP 2014), The Hague, The Netherlands, 6–11 April 2014; pp. 2002–2004.
31. Lopez, D.; Coves, A.; Bronchalo, E.; Torregrosa, G.; Bozzi, M. Practical Design of a Band-Pass Filter using EBG SIW Technology. In Proceedings of the 48th Microwave European Conference, Madrid, Spain, 23–27 September 2018; pp. 77–80.
32. Li, D.; Tong, C.; Bao, J.; Peng, P.; Yu, D. A novel bandpass filter of substrate integrated waveguide (SIW) based on S-shaped EBG. *Prog. Electromagn. Res. Lett.* **2013**, *36*, 191–200. [CrossRef]
33. Yu, M.-X.; Shi, Z.-Z. A novel millimetre wave EBG-SIW wideband delay line. *Int. J. Electron.* **2013**, *100*, 762–769. [CrossRef]
34. Raman, S.; Rucky, F.; Rebeiz, G. A High-Performance W-Band Uniplanar Subharmonic Mixer *IEEE Trans. Microw. Theory Tech.* **1997**, *45*, 955–959. [CrossRef]
35. Kim, J.; Qian, Y.; Feng, G.; Ma, P.; Judy, J.; Chang, M.F.; Itoh, T. A Novel Low-Loss Low-Crosstalk Interconnect for Broad-Band Mixed-Signal Silicon MMIC's. *IEEE Trans. Microw. Theory Tech.* **1999**, *47*, 1830–1834.
36. Dancila, D.; Rottenberg, X.; Tilmans, H.A.C.; De Aedt, W.R.; Huynen, I. 57-64 GHz Seven-Pole Bandpass Filter Substrate Integrated Waveguide (SIW) in LTCC. In Proceedings of the 2011 IEEE MTT-S International Microwave Workshop on Millimeter Wave Integration Technologies, Sitges, Spain, 15–16 September 2011; pp. 200–203.
37. Hizan, H.M.; Ambak, Z.; Ibrahim, A.; Yusoff, M.Z.M.; Kanesan, T. Effect of Insertion Losses on Millimeter-Wave SIW Filters Using LTCC Technology. In Proceedings of the 2015 IEEE International RF and Microwave Conference (RFM 2015), Kuching, Malaysia, 14–16 December 2015; pp. 64–68.
38. Bertrand, M.; Rehder, G.P.; Serrano, A.L.C.; Gomes, L.G.; Pinheiro, J.M.; Alvarenga, R.C.A.; Kabbani, N.; Kaddour, D.; Puyal, V.; Ferrari, P. Integrated Waveguides in Nanoporous Alumina Membrane for Millimeter-Wave Interposer. *IEEE Microw. Wirel. Compon. Lett.* **2019**, *29*, 83–87. [CrossRef]
39. Hamoir, G. Microwave Devices Based on Unbiased Tunable Ferromagnetic Nanowire Arrays. Ph.D. Thesis, Université catholique de Louvain, Ottignies-Louvain-la-Neuve, Belgium, 2014.
40. Lasers, O. *J-1064/355 System—Operation Manual*, Technical Report; Oxford Lasers Ltd.: Didcot, Oxon, UK, 2010.
41. Emplit, A.; Tooten, E.; Xhurdebise, V.; Huynen, I. Multifunctional material structures based on laser etched carbon nanotubes arrays. *Micromachines* **2014**, *5*, 756–765. [CrossRef]
42. Eul, H.; Schiek, B. Reducing the Number of Calibration Standards for Network Analyzer Calibration. *IEEE Trans. Instrum. Meas.* **1991**, *40*, 732–735. [CrossRef]

Article

Design and Performance of a J Band MEMS Switch

Naibo Zhang [1,*], Ze Yan [2], Ruiliang Song [1], Chunting Wang [1], Qiuquan Guo [3,*] and Jun Yang [3]

1 The 54th Research Institute of China Electronic Science and Technology Group Corporation, Beijing 100070, China
2 Beihang University, Beijing 100083, China
3 Department of Mechanical & Materials Engineering, University of Western Ontario, London, ON N6A 3K7, Canada
* Correspondence: zhangnaibai@163.com (N.Z.); Qguo29@uwo.ca (Q.G.)

Received: 14 June 2019; Accepted: 12 July 2019; Published: 13 July 2019

Abstract: This paper presents a novel J band (220–325 GHz) MEMS switch design. The equivalent circuits, the major parameters, capacitance, inductance and resistance in the circuit were extracted and calculated quantitatively to carry out the radio frequency analysis. In addition, the mechanical property of the switch structure is analyzed, and the switching voltage is obtained. With the designed parameters, the MEMS switch is fabricated. The measurement results are in good agreement with simulation results, and the switch is actuated under a voltage of ~30 V. More importantly, the switch has achieved a low insertion loss of ~1.2 dB at 220 GHz and <~4 dB from 220 GHz to 270 GHz in the "UP" state, and isolation of ~16 dB from 220 GHz to 320 GHz in the "DOWN" state. Such switch shows great potential in the integration for terahertz components.

Keywords: J band; MEMS; switch

---

1. Introduction

There is a growing interest in terahertz components in the frequency from 100 GHz to 10 THz for potential applications in security scanning, atmospheric monitoring, medical imaging and ultrafast wireless communications. Silicon-based terahertz monolithic integrated circuits for communication circuits and radar systems have been developed with high integration and low cost as the main advantages [1]. With the growing requirement for multi-function and reconfigurability of THz circuits and systems, the tunable devices have become necessary for research, in which switches are the most basic components. At the same time, the switch is also the key component to control the high frequency signal in the communication system [1–3]. However, semiconductor switches are not suitable for terahertz frequency applications due to insertion loss, power handling capacity and linearity limitations.

RF micro-electromechanical (MEMS) switches have demonstrated excellent linearity, high isolation, low insertion loss and low power consumption [3]. Previous studies on MEMS switches have focused on frequencies lower than 100 GHz and have achieved good performances [4,5]. In comparison, RF-MEMS switches can be integrated with coplanar waveguide (CPW) transmission lines and together with other RF devices. In recent research, THz communication systems are already being developed. To meet this requirement of terahertz system, a wideband RF-MEMS switch using BiCMOS process was firstly presented in [1]. It realized a return loss of 24 to 12 dB and an insertion loss of 1.2 to 2.7 dB within the frequency range 180–250 GHz. After that, other RF-MEMS switches fabricated on chip were also proposed [2,3]. On the other hand, MEMS-based waveguide switches for terahertz band were demonstrated in [6,7]. Most recently, RF-MEMS waveguide switch has achieved an isolation of 19–24 dB and insertion loss of 2.5–3 dB for 500–750 GHz, respectively. In a recent publication, the authors have studied these MEMS switches in the frequency band between 500 and 750 GHz [3,6,7], which was the highest frequency RF MEMS device reported so far, with an insertion loss of 2–4 dB and DC driving voltage of 30–60 V.

This paper introduces a novel J Band (220 GHz–325 GHz) MEMS switch with operating frequency from 220 GHz to 320 GHz. The equivalent circuits are discussed and extracted for the switch, and the parameters in the circuit are calculated quantitatively. The measurement results show a good agreement with simulation results, and the insertion loss is ~−1.2 dB in the "UP" state when the frequency is 220 GHz, and the isolation is better than 16 dB across the band in the "DOWN" state. The switch is actuated under a voltage of ~30 V.

2. Design

The MEMS switch is designed as Figure 1a,b; the switch consisted of a CPW (coplanar waveguide) transmission line, DC bias and switch beam. By controlling the DC bias voltage, the switch beam can be brought into contact with the CPW section to switch into the "DOWN" state. To realize a switch working at THz frequencies, there are three major challenges: low insertion loss in the "UP" state, high isolation in the "DOWN" state, and robust mechanical properties. Reducing the capacitance of the switch to the ground can reduce the insertion loss of the switch in the "UP" state, so the ground structure between signal line is designed as the comb-like structure in Figure 1a. A driving voltage electrode plate is designed to complement the ground comb, which will increase the inductance value and thus increase the isolation of the signal to the ground. MEMS switch beam directly connecting the signal lines can make the switch inductance smaller in the "UP" state, which can reduce the reflected signal energy. In order to improve the isolation in the "DOWN" state, the capacitance value in the "DOWN" state needs to be increased, which is realized by covering a thin isolation layer Si_3N_4 on the bottom plate of the switch with a thickness of ~0.1 µm.

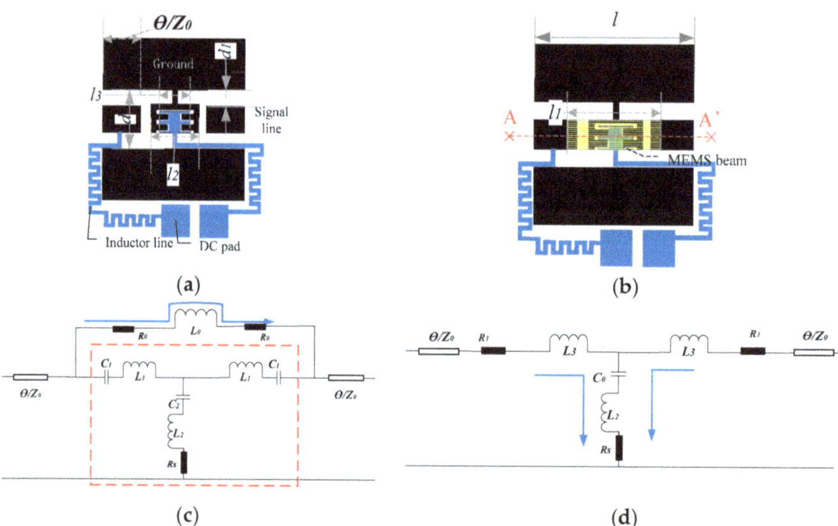

Figure 1. THz RF micro-electromechanical (MEMS) switch. (**a**) Switch structure without beam, $d = 60$ µm, $d_1 = 18$ µm, $l_2 = 60$ µm, $l_3 = 40$ µm. (**b**) Switch structure with beam, $l = 204$ µm, $l_1 = 134$ µm. (**c**) The equivalent circuit of structure when the switch is in the "UP" state. (**d**) The equivalent circuit of structure when the switch is in the "DOWN" state.

2.1. Radio Frequency Analysis

The equivalent circuits are analyzed in order to achieve a low insertion loss in the "UP" state and a high isolation in the "DOWN" state. The equivalent circuits of structure in the "UP" and "DOWN" states are shown in Figure 1c,d. The inductors L_0, $R_{0/1}$ are generated by MEMS switch beam, C_1 is the capacitor produced by the gap between signal line and ground, L_1, L_2 are produced by ground between two signal line, and C_2 is the capacitor between switch beam and ground.

From the equivalent circuits, we can determine that there are three parameters to be quantitatively calculated: resistance (R), inductance (L) and capacitance (C). The inductance and capacitance can be calculated by formulas in literatures [8], and the resistance [9] is calculated as following.

Current distribution is not uniform when the device is working at high frequencies due to skin effect in metal film; a diagram of the current distribution profile is illustrated in Figure 2. Due to the skin effect of current, the current mainly distributes from r to r_0, and the current density is expressed as,

$$J = J_0 e^{-\alpha(r_0 - r)},\ \alpha = \sqrt{\frac{w\mu\sigma}{2}} \tag{1}$$

where, J_0 is surface current density, α is attenuation constant, r_0 is the outer length of current distribution, r is the inner length of current distribution, w is frequency, σ is conductivity (= 4.1 × 10^7 S/m), μ is magnetic permeability, and the skin depth of current is given as,

$$\Delta = \sqrt{\frac{2}{w\mu\sigma}} \tag{2}$$

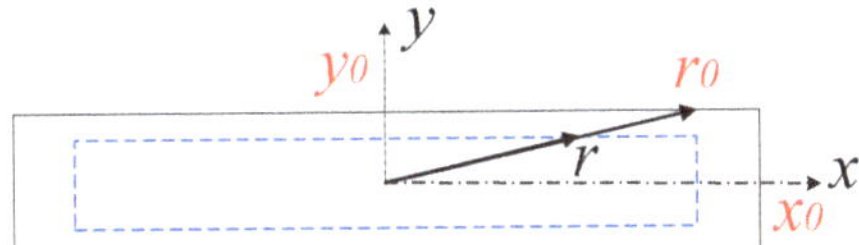

Figure 2. Side view of current distribution in metal film.

The current consists of the current density integral of the whole plane,

$$I = \iint J_0 (e^{-\alpha(x_0 - x)} + e^{-\alpha(y_0 - y)})dS \approx \frac{1}{\sqrt{2}}\pi\sigma E_0 \frac{1}{\alpha}\sqrt{x_0{}^2 + y_0{}^2} = \frac{1}{\sqrt{2}}\pi\sigma E_0 \Delta\sqrt{x_0{}^2 + y_0{}^2} \tag{3}$$

where, x_0 and y_0 are border lengths, E_0 is the equivalent electrostatic field.

$$R = \frac{E_0 l}{I} = \frac{l\sqrt{2}}{\pi\sigma\Delta\sqrt{x_0{}^2 + y_0{}^2}} = \frac{1}{2 \times 10^4 \times \sqrt{x_0{}^2 + y_0{}^2}} \tag{4}$$

where, l is unit length, the parameters are designed as, x_0 = 12 μm, and y_0 = 0.5 μm.

In order to obtain good radio frequency performances, including low insertion loss in the "UP" state and high isolation in the "DOWN" state, it can be obtained by calculation that when the switch is in the "UP" state, its equivalent circuit parameters are: R_0 = 2.4 Ω, L_0 = 24 pH, C_1 = 6.7 fF, L_1 = 19 pH, C_2 = 5 fF, L_2 = 6.5 pH, and R_s = 2.5 Ω, respectively. When the switch is in the "DOWN" state, the equivalent circuit parameters are: R_1 = 2.4 Ω, L_3 = 23 pH, C_0 = 134 fF, L_2 = 6.3 pH, and R_s = 2.5 Ω, respectively.

The parameters of the switch structure can be calculated from the capacitance, inductance and resistance calculation formulas, where the capacitance and inductance calculation formulas obtained from the formula in literature [8], and the resistance is calculated from Formula (4). According to the equivalent circuit, calculated parameters and optimization, the parameters of the MEMS switch can be obtained as follows. The switch dimensions in Figure 1 are as, Z_0 = 50 Ω, d = 60 μm, d_1 = 18 μm, l_2 = 60 μm, l_3 = 40 μm, l = 204 μm, and l_1 = 134 μm, respectively.

The calculated results based on calculation parameters are compared with those of switch structure simulated based on finite element method, which are shown in Figure 3, and the frequency range is from 220 GHz to 320 GHz. When the MEMS switch is in the "UP" state, the insertion loss is ~−2.5 dB, and both two curves of simulation and calculation results have the same trend of change. When the MEMS switch is in the "DOWN" state, the isolation is ~16 dB, while two curves have some differences. The main reason for the deviation between two curves in Figure 3a,b is that the capacitance and inductance produced by the comb structure of the bottom electrode in the switch are small, which are neglected in the equivalent circuit, while these parameters are all considered in the simulation of the equivalent circuit in terahertz band. The difference in Figure 3d is because the capacitance and inductance between signal line and ground is ignored, which leads to ~2 dB differences when frequency is from 280 GHz to 320 GHz. The MEMS switch structures simulations are carried out by Ansoft HFSS, and circuits simulations are carried out by ADS.

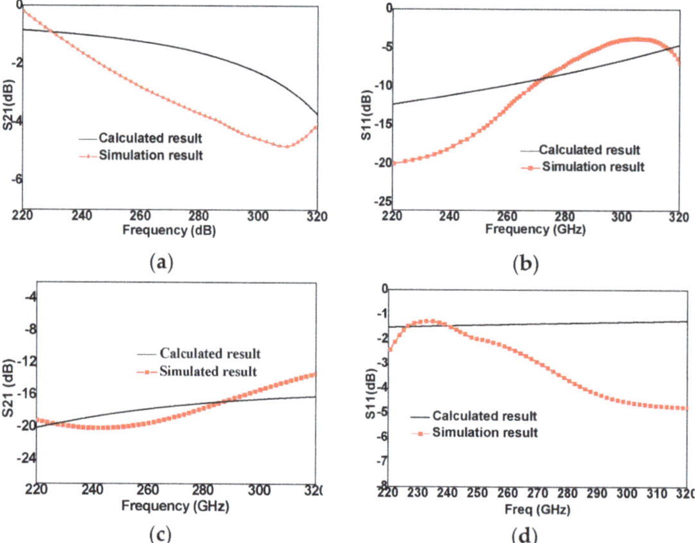

Figure 3. Calculated circuit and simulated S-parameters of MEMS switch. (**a**) Insertion loss in the "UP" state. (**b**) Return loss in the "UP" state. (**c**) Isolation in the "DOWN" state. (**d**) Return loss in the "DOWN" state.

2.2. Analysis of Mechanical Properties

The MEMS switch is a symmetrical structure. Figure 4a shows the side view of the switch, the electrode plate is from $(l_1 - x)$ to x, and the electrostatic force is loaded in the middle part of the beam, with the whole size of the beam as l_1. ξ is the pressure. The elastic coefficient (k) of MEMS switch is in two parts, k_1 and k_2.

(a) The elastic coefficient k_1 is due to the stiffness of the bridge, which accounts for the material characteristics such as Young's modulus [10], E (Pa), and the moment of inertia, I (m^4) = $l_1 t^2/12$, t is the thickness of switch beam. The deflection (y_1) found by evaluating the integral is,

$$y_1 = \frac{1}{EI}\int_{l_1-x}^{x} \frac{\xi}{48}({l_1}^3 - 6{l_1}^2 a + 9 l_1 a^2 - 4a^3)da \tag{5}$$

Elastic coefficient (k_1) caused by Stiffness of Beams is,

$$k_1 = -\frac{\xi l_1}{y_1} = 32Ew(\frac{t}{l_1})^3 \frac{1}{8(x/l_1)^3 - 20(x/l_1)^2 + 14(x/l_1) - 1} \tag{6}$$

(b) The elastic coefficient k_2 is due to the biaxial residual stress, $S'(\text{Pa}) = \sigma(1 - \nu)l_1 t$, where ν is the Poisson's ratio [8]. The deflection (y_2) found by evaluating the integral is,

$$y_2 = -\int_{l_1 - x}^{x} \frac{\xi}{2S}(l_1 - a)da \tag{7}$$

Elastic coefficients induced by biaxial residual stresses is,

$$k_2 = -\frac{\xi l_1}{y_2} = 8\sigma(1-\nu)w(\frac{t}{l_1})\frac{1}{3 - 2(x/l_1)} \tag{8}$$

Figure 4b is the MEMS structure beam, and holes on the beam surface are designed. Photoresist was used as sacrificial material and removed to fabricate the through holes. The hole diameter $r = 5$ μm, and the distance $d = 8$ μm; eighteen holes are designed in two rows on the beam surface. The holes will bring certain elastic coefficient error which has been well characterized in the literature [11]. The holes are helpful to release some of the residual stress in the beam but will lead to the reducing of the Young's modulus [12]. The reduction of the residual stress is equivalent to $\sigma = (1 - \mu)\sigma_0$, where σ_0 is the residual stress with no holes. The elastic coefficient is revised to $k = \lambda_1(k_1 + k_2)$, $(0 < \lambda_1 < 1)$.

The total elastic coefficient is,

$$k = \lambda(k_1 + k_2) = 32Ew(\frac{h_3}{l_1})^3 \frac{\lambda}{8(x/l_1)^3 - 20(x/l_1)^2 + 14(x/l_1) - 1} + 8\sigma(1-\nu)w(\frac{h_3}{l_1})\frac{\lambda}{3 - 2(x/l_1)} \tag{9}$$

Through analysis and calculation, the parameters in this equation λ, σ, E, t, and k are 0.75, 56 MPa [13], 46 GPa [13], 1 μm, and 5.9 N/m, respectively. The controlled DC voltage is represented as $V = t_1[2k{\cdot}\Delta d/(\varepsilon{\cdot}\varepsilon_0{\cdot}S)]^{0.5}$ [8], $\varepsilon_0 = 8.85 \times 10^{12}$ F/m, $\varepsilon = 1$, Δd is the displacement of switch beam, S is the area of electrodes, and k is the elastic coefficient. The calculated DC voltage is ~35 V.

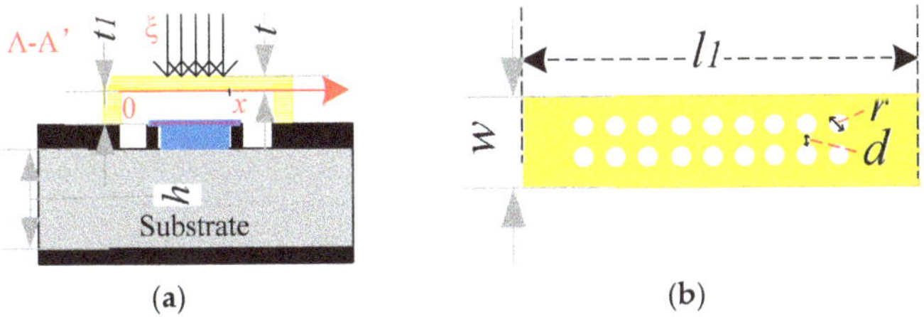

Figure 4. (**a**) The side view of switch, $t = 1$ μm, $t_1 = 1$ μm, $t_2 = 1$ μm. (**b**) The MEMS switch beam.

3. Results

In order to verify the switch design, the switches were fabricated, and the performances were measured. An image of the completed MEMS switch is provided in Figure 5a, and a two-port on wafer measurement is set up as shown in Figure 5b. R&S ZVA 40 vector network analyzer and VNA extender were used for testing the S ($S21$, $S11$) parameters, and the CASCADE probe station was used for measurement. All experiments were performed under ambient environment without any packaging. A G-S-G probe with 100 μm spacing was used for MEMS switch testing, and a through-reflect-line (TRL) calibration was applied using WinCal XE calibration software. The "Line" standards are used for higher calibration accuracy [3]. Calibrated measurement shows the loss of the 50 Ω CPW is around ~1.8 dB/mm at 200 GHz and around ~2.5 dB/mm at 300 GHz, which is consistent with the simulated results.

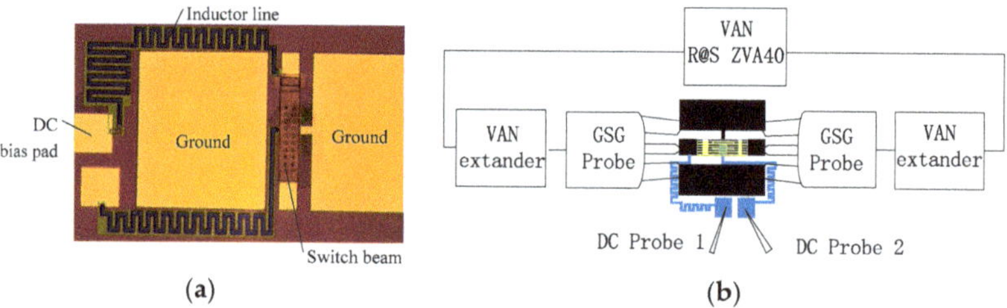

Figure 5. (**a**) Fabricated THz MEMS switch. (**b**) Set up of the two-port on wafer measurement.

Figure 6 shows the RF performance of MEMS switch, and the frequency is from 220 GHz to 320 GHz. The black color curves are simulation results and red color curves are experimental results. The experimental results show good agreement with simulation results. When MEMS switch is in the "UP" state, the insertion loss is ~−1.2 dB at 220 GHz, and is <−4 dB from 220 GHz to 270 GHz. When MEMS switch is in the "DOWN" state, the isolation is ~<−16 dB from 220 GHz to 320 GHz. The DC voltage is ~30 V, which is less than calculated result (35 V), the reason is that the surface of switch beam is reduced and the hole on the switch is enlarged, which directly leads to the decrease of the elasticity coefficient of the switch, thus, the driving voltage of the switch being reduced. In addition, the out-of-plane deformation of a cantilever beam with residual stress gradient will also reduce the driving voltage [14,15].

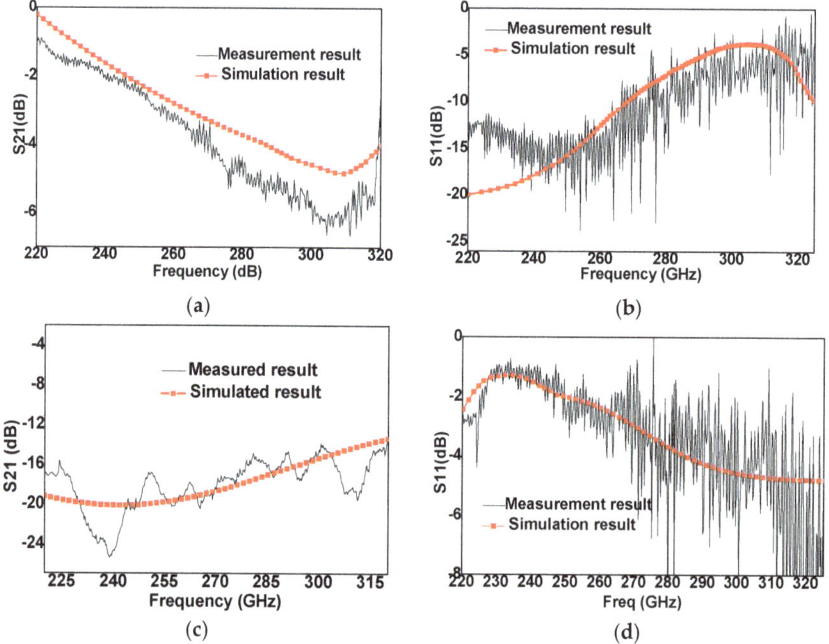

Figure 6. Measured and simulated results. (**a**) Insertion loss at "UP" state. (**b**) Return loss at "UP" state. (**c**) Isolation at "DOWN" state. (**d**) Return loss at "DOWN" state.

4. Discussion

When the MEMS structure have been determined, the insertion loss is mainly influenced by two parts, (1) loss caused by the substrate dielectric constant and the substrate thickness, which is ~200 μm in this study. In order to reduce the insertion loss, the thickness of the substrate can be further reduced, such as 50 μm; The influence of the parameters of the switch on the switch RF performance is analyzed in two aspects: Firstly, circuit and theoretical analysis. Secondly, optimization simulation by simulation software. In this paper, related work has been undertaken in the early stage. The structure parameters of the switch are obtained through the analysis of the above two aspects. In the design, the thickness of the substrate is not the optimal value, but it is limited by the previous process, so the thickness of the substrate has not been reduced. (2) Loss caused by the fabrication process; there are two factors, fabrication tolerance and surface roughness. The deviation of the structure size often leads to impedance mismatch, which the surface roughness produced by fabrication limit often leads to increasing loss. So, in addition to the structural design, the loss caused by the thickness of the substrate and the loss caused by the fabrication process should be considered for the practical application use. Almost 10 samples were tested, and the results were almost the same. A discrepancy between simulation and measurement results is observed; a large fluctuation exists in the experimental data of Figure 6c, and a low agreement appears in Figure 6b (low frequency) and Figure 6d (high frequency). The reasons are as follows: (1) The parasitic parameters, such as inductance and capacitance, produced by the smoothness of the switch film surface will affect the RF performance parameters of the switch. (2) Ansof HFSS is used to simulate the RF performance of the switch. The size of the peripheral air box is set at the designing, but no metal package is added during the test, which will also affect the RF performance, resulting in some inconsistencies between the simulation and test results [16,17]. Furthermore, the main reasons for the uncertainty information about the experimental results is as follows: (1) Uncertainties from fabrication. Fabrication errors have great impact on RF performance, mainly due to the height of anchor zone and film thickness. The height of switch gap (anchor height) and film thickness designed here are 1 μm. When the height of switch gap changes, the RF performance of switch will be greatly affected. In addition, the size of the hole has an impact on the performance. When the diameter of the hole is less than (3–4)t_1 and the switch is in the "UP" state, the influence of the hole is neglected. This is because the edge effect makes the hole "filled", but when it is in "DOWN" state, the capacitance will be affected, where the influence ratio is the proportion of the film area reduced. At the same time, the fabrication will bring over-corrosion effect, the hole diameter will be larger than the designed diameter, which will affect the "UP" capacitor and further affect the radio frequency performance. (2) Uncertainties in testing, such as test instruments, test operation, etc. During testing, the contact between probe and switches will affect the test results. When the contact is insufficient, the contact resistance and inductance will increase, resulting in deviation of test results, such as resonance frequency offset. In addition, when the test transmission line interface is not fully connected or the transmission line jitters, there will be an impact on the test results.

To compare the proposed THz MEMS switch in this study and reported THz switch, Table 1 shows recently published THz MEMS switches. The proposed switch with a low insertion loss which is comparable to the switch performances [1,2,6], and low DC driving voltage [1–3,6,7]. In other words, the proposed switch can have both the advantages of a low DC driving voltage, a low insertion loss and high return loss of passband. Compared with the switches with the same center frequency (220 GHz) in literature [1], the switch in this study shows a much better performances than performances in literature [1]. When compared with literatures [3,6,7], the proposed switch in this study shows a lower DC driving voltage. These advantage features of proposed switch will be useful for many system applications, such as, MEMS dives and THz communication system.

Table 1. Comparison of different THz switches.

Title 1	Frequency (GHz)	S21 (dB)	S11 (dB)	DC Driving Voltage (V)
Ref. [1]	170–220	1.9 dB@220 GHz	−12	50
Ref. [2]	110–170	−1.5	−19	60
Ref. [3]	500–750	2.7 dB@600 GHz	−15	60
Ref. [6]	330–500	−5	−20	95
Ref. [7]	500–750	−3.5	−16	36
This work	220–3200	1.2 dB@220 GHz	−16	30

5. Conclusions

A novel J band MEMS switch is designed, analyzed, fabricated, and measured. A low insertion loss in the "UP" state and high isolation in the "DOWN" state were achieved. The equivalent circuits retaining all the components were used and the switch parameters in the circuit were calculated quantitatively. The MEMS switch was fabricated, and the completed device was actuated under a voltage of ~30 V. The experimental results matched expectation and fit well with the simulation results. A switch with the insertion loss of −1.2 dB in "UP" state when the frequency is 220 GHz, and isolation better than 16 dB across the band in the "DOWN" state was fabricated.

Author Contributions: N.Z. is responsible for theoretical analysis, mathematical modeling, structure and circuit simulation, parameter extraction, paper writing, etc. Z.Y. is responsible for literatures retrieval, language modification, etc. R.S., C.W., Q.G. and J.Y. are responsible for language modification and formula parameter calculation, etc.

Funding: This research was funded by National Natural Science Foundation of China, grant number 61504124.

Conflicts of Interest: The authors declare no conflict of interest

References

1. Du, Y.J.; Su, W.; Tolunay, S.; Zhang, L.; Kaynak, M.; Scholz, R.; Xiong, Y.Z. 220GHz wide-band MEMS switch in standard BiCMOS technology. In Proceedings of the 2015 IEEE Asian Solid-State Circuits Conference (A-SSCC), Xiamen, China, 9–11 November 2015; pp. 1–4.
2. Wipf, S.T.; Göritz, A.; Wietstruck, M.; Wipf, C.; Tillack, B.; Kaynak, M. D-Band RF-MEMS SPDT Switch in a 0.13 μm SiGe BiCMOS Technology. *IEEE Microw. Wirel. Compon. Lett.* **2016**, *26*, 1002–1004. [CrossRef]
3. Feng, Y.; Barker, N.S. High performance 500–750 GHz RF MEMS switch. In Proceedings of the 2017 IEEE MTT-S International Microwave Symposium (IMS), Honolulu, HI, USA, 4–9 June 2017; pp. 1095–1097.
4. Sudhanshu, S.; Vinoy, K.J.; Ananthasuresh, G.K. Surface-Micromachined Capacitive RF Switches with Low Actuation Voltage and Steady Contact. *J. Microelectromech. Syst.* **2017**, *26*, 643–652.
5. Persano, A.; Quaranta, F.; Capoccia, G.; Proietti, E.; Lucibello, A.; Marcelli, R.; Bagolini, A.; Iannacci, J.; Taurino, A.; Siciliano, P. Influence of design and fabrication on RF performance of capacitive RF MEMS switches. *Microsyst. Technol.* **2016**, *22*, 1741–1746. [CrossRef]
6. Laemmle, D.; Schlaak, H.F.; Weickhmann, C.; Jakoby, R. Proof of concept for a WR-2.2 MEMS waveguide switch. In Proceedings of the 2016 41st International Conference on Infrared, Millimeter, and Terahertz waves (IRMMW-THz), Copenhagen, Denmark, 25–30 September 2016; pp. 1–2.
7. Shah, U.; Reck, T.; Frid, H.; Jung-Kubiak, C.; Chattopadhyay, G.; Mehdi, I.; Oberhammer, J. A 500–750 GHz RF MEMS Waveguide Switch. *IEEE Trans. Terahertz Sci. Technol.* **2017**, *7*, 326–334. [CrossRef]
8. Rebeiz, G.M. *RF MEMS: Theory, Design, and Technology*; John Wiley & Sons: Hoboken, NJ, USA, 2004.
9. Zhang, N.; Wang, C.; Lang, L.; Huang, J. A Fixed Miniaturization 90° Phase Shifter for VHF-Band High-Power Applications. *IEEE Access* **2019**, *7*, 34395–34402. [CrossRef]
10. Roark, R.J.; Young, W.C. *Formulas for Stress and Strain*, 6th ed.; McGraw-Hill: New York, NY, USA, 1989.
11. Zhang, N.; Deng, Z. CPW Tunable Band-stop Filter Using Hybrid Resonator and Employing RF MEMS Capacitors. *IEEE Trans. Electron. Dev.* **2013**, *60*, 2648–2655. [CrossRef]

12. Rabinovich, V.L.; Gupta, R.K.; Senturia, S.D. The effect of release-etch holes on the electromechanical behaviour of MEMS structures. In Proceedings of the International Solid State Sensors and Actuators Conference (Transducers '97), Chicago, IL, USA, 19 June 1997; Volume 2, pp. 1125–1128.
13. WebElements. Available online: http://www.webelements.com/ (accessed on 5 October 2018).
14. Persano, A.; Iannacci, J.; Siciliano, P.; Quaranta, F. Out-of-plane deformation and pull-in voltage of cantilevers with residual stress gradient: Experiment and modelling. *Microsyst. Technol.* **2018**. [CrossRef]
15. Ballestra, A.; Brusa, E.; De Pasquale, G.; Munteanu, M.G.; Soma, A. FEM modelling and experimental characterization of microbeams in presence of residual stress. *Analog Integr. Circuits Signal Process.* **2010**, *63*, 477–488. [CrossRef]
16. Persano, A.; Siciliano, P.; Quaranta, F.; Taurino, A.; Lucibello, A.; Marcelli, R.; Capoccia, G.; Proietti, E.; Bagolini, A.; Iannacci, J. Wafer-level micropackaging in thin film technology for RF MEMS applications. *Microsyst. Technol.* **2018**, *24*, 575–585. [CrossRef]
17. Leedy, K.D.; Strawser, R.E.; Cortez, R.; Ebel, J.L. Thin-Film Encapsulated RF MEMS Switches. *J. Microelectromech. Syst.* **2017**, *16*, 304–309. [CrossRef]

Article

A Stripline-Based Planar Wideband Feed for High-Gain Antennas with Partially Reflecting Superstructure

Affan A Baba [1,*], Raheel M Hashmi [1], Mohsen Asadnia [1], Ladislau Matekovits [2] and Karu P Esselle [1]

1 School of Engineering, Macquarie University, Sydney, New South Wales 2109, Australia; raheel.hashmi@mq.edu.au (R.M.H.); mohsen.asadnia@mq.edu.au (M.A.); karu.esselle@mq.edu.au (K.P.E.)
2 Department of Electronics and Telecommunication, Politecnico di Torino, 10129 Turin, Italy; ladislau.matekovits@polito.it
* Correspondence: affan.baba@mq.edu.au

Received: 12 April 2019; Accepted: 5 May 2019; Published: 7 May 2019

Abstract: This paper presents a new planar feeding structure for wideband resonant-cavity antennas (RCAs). The feeding structure consists of two stacked dielectric slabs with an air-gap in between. A U-shaped slot, etched in the top metal-cladding over the upper dielectric slab, is fed by a planar stripline printed on the back side of the dielectric slab. The lower dielectric slab backed by a ground plane, is used to reduce back radiation. To validate the wideband performance of the new structure, in an RCA configuration, it was integrated with a wideband all-dielectric single-layer partially reflecting superstructure (PRS) with a transverse permittivity gradient (TPG). The single-layer RCA fed by the U-slot feeding structure demonstrated a peak directivity of 18.5 dBi with a 3 dB directivity bandwidth of 32%. An RCA prototype was fabricated and experimental results are presented.

Keywords: high-gain; compact; wideband; resonant cavity; Fabry–Perot cavity; cavity resonator; EBG resonator

1. Introduction

Common approaches to achieve high antenna directivity include the use of antenna arrays, which require a complex feeding network, or reflector antennas, which are bulky in nature. Resonant-Cavity Antennas (RCAs) have been investigated extensively as a convenient alternative in applications requiring directivity in the range of 15–25 dBi [1–4]. An RCA is formed by placing a partially reflecting superstructure (PRS) in front of a primary source antenna (such as a slot, or dipole), above a metallic ground plane or Artificial Magnetic Conductor (AMC) [5]. The directivity of this source antenna is significantly increased by multiple reflections between the PRS and the ground plane. The PRS can take various forms, depending on the design approach and fabrication methods used to construct the RCAs, including 3-D Electromagnetic Band Gap (EBG) structures [6]; 2-D metallo-dielectric frequency-selective surfaces [7]; or stacks of unprinted dielectric slabs [8]. Thus, RCAs are also referred to as EBG resonator antennas [9], Fabry–Perot cavity antennas [10], and 2D leaky-wave antennas [11].

A number of wideband RCAs having all-dielectric PRSs have been proposed with peak gain and directivity greater than 15 dBi and 3 dB directivity bandwidths that are significantly greater than the initial RCAs (1–3%) [12,13]. However, with the increase in bandwidth, the input matching becomes increasingly challenging. This effect is more pronounced when the peak gain of the RCAs ranges between 15–20 dBi, which is required in most medium-range applications. Therefore, rectangular slots cut in a ground plane, fed by waveguide-to-SMA adaptors (shown in Figure 1a) are commonly used to feed wideband high-gain RCAs, particularly when the bandwidths exceed beyond 20% [14–19].

However, standard waveguide adaptors are quite expensive, and mechanically bulky. Although they present a reasonable feeding method for antenna testing and characterization, their use adds significant cost, weight, and volume in most practical RCAs.

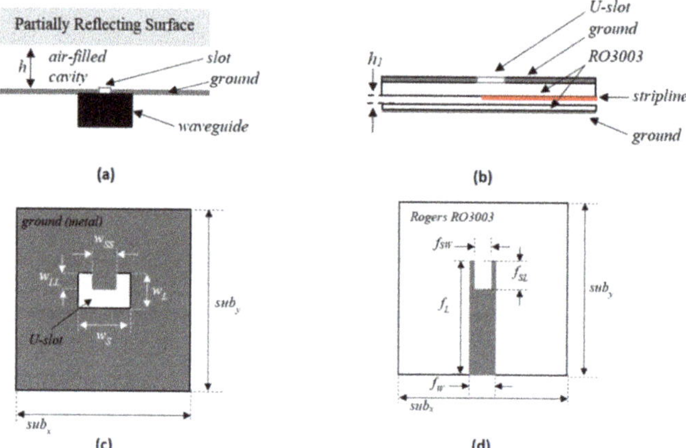

Figure 1. Structural schematic of the proposed planar feed, (**a**) configuration of wideband resonant-cavity antennas (RCAs) fed using conventional waveguides, (**b**) vertical cross section of the antenna, showing the stripline and the air-gap, (**c**) top view of the antenna, showing the U-slot cut in the ground plane, (**d**) horizontal cross section of the antenna, showing the stripline.

In this paper, a planar, wideband feed antenna is proposed, which can be used for feeding wideband RCAs. It can easily be fabricated through the standard-printed-circuit board (PCB) manufacturing process, and can easily be interfaced with any type of wideband PRS, to form a stand-alone RCA. Initially, the planar wideband feed antenna itself was numerically designed and studied using the Computer Simulation Technology Microwave Studio (CST MWS) time-domain solver. Once the desired performance was obtained, it was interfaced with a wideband dielectric PRS developed in Reference [16] and an RCA was designed using the CST-MWS time-domain solver. The RCA fed by this planar feed demonstrated a peak gain of 18.5 dBi with a 3 dB gain-bandwidth of 32%. A prototype was fabricated and measured to validate the performance

2. Design of Planar Wideband Feed Antenna

Figure 1b–d show the schematic of the proposed planar, wideband feed antenna. It consists of two substrates in a stacked configuration, with an air-gap in between (Figure 1b). The permittivity of the substrates is $\varepsilon_r = 3.02$ and the loss tangent is tan $\delta = 0.0019$. The top side of the upper substrate is metal-cladded, with a U-slot feed etched in the center (see Figure 1c). A stripline is etched on the bottom side of the upper substrate with a U-shaped stub added at the coupling end (see Figure 1d). The lower substrate has no metal cladding on the top side, but is backed by fully-metallic cladding. The lower substrate is placed at a distance h_1 away from the stripline, and circumvents back radiation, as shown in later sections. All the design parameters shown in Figure 1 are given in Table 1.

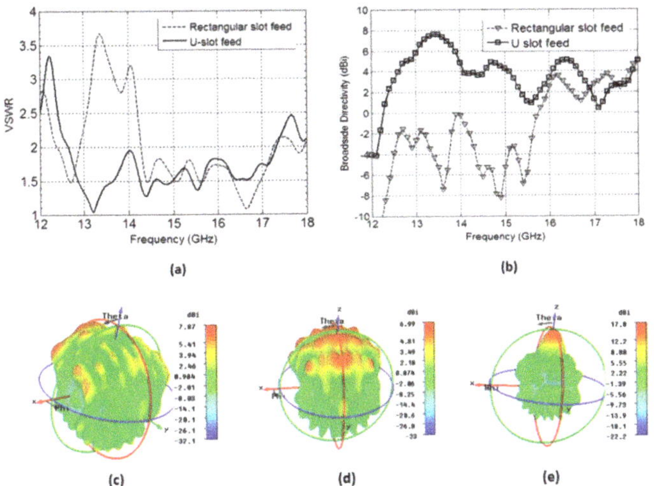

Figure 2. Comparison of rectangular slot with U-slot, when fed by the stripline, (**a**) voltage standing wave ratio (VSWR) of the rectangular slot and U-slot antenna, (**b**) broadside directivity of the rectangular slot and U-slot antenna, (**c**) radiation patterns of rectangular slot antenna and (**d**) radiation patterns of U-slot antenna (**e**) radiation patterns of the RCA with U-slot antenna at 15 GHz.

Table 1. Parameter values for the stripline feed structure (in mm).

Parameter	w_S	w_{SS}	w_L	w_{LL}	f_W
	9.0	4.5	13.125	2.875	6
Parameter	f_{SL}	f_{SW}	sub_x	sub_y	f_L
	6.0	3.0	80.0	80.0	47.5

The RCA fed by a waveguide-to-SubMiniature version A (SMA) transition in [16] covers the entire Ku-band, with a peak gain of 20.7 dBi. Therefore, we selected this frequency band as a reference to replace the waveguide-to-SMA transition with a planar feed. Initially, we began with a simple rectangular slot fed by a stripline. The length and width of the rectangular slot were 13.125 mm and 9 mm, respectively. This configuration provided a VSWR 2:1 bandwidth extending from 14.2 GHz–17.3 GHz, providing matching in the upper Ku-band (see Figure 2a). Another important aspect to consider is that a rectangular slot fed by a waveguide-to-SMA adaptor, as done in Reference [16], provides a broadside-directed radiation pattern over the entire matched bandwidth. This characteristic is crucial to achieving high gain from an RCA. The rectangular slot fed by a stripline, however, does not radiate towards the broadside over the entire VSWR 2:1 bandwidth, as shown in Figure 2b,c. To improve the coupling towards the broadside, the rectangular slot was modified to form a U-slot, which directed the maximum radiation towards the broadside (see Figure 2b,d), and improved the VSWR 2:1 bandwidth to extend from 12.4 GHz to 18.2 GHz, which covers nearly the entire Ku-band. Extensive parametric studies were carried out to tune the performance of the U-slot and the stripline to achieve the optimal performance, but only the key results are presented and discussed here for brevity. The final values of the parameters that provide the best compromise between VSWR 2:1 bandwidth and broadside-directed radiation are shown in Table 1.

3. Resonant-Cavity Antennas (RCA) Design and Interfacing with Planar Feed

In this section, we describe the interfacing of the planar feed with a wideband dielectric PRS to design an RCA. The schematic of the wideband dielectric PRS is shown in Figure 3. The PRS has a circular shape and a diameter D = 78 mm ($2.6\lambda_0$ at 10 GHz) and is placed at a distance h = 16 mm ($\sim 0.53\lambda_0$) from the top part of the feed antenna, as shown in Figure 3. In this configuration, the metallic cladding on the top side of the upper substrate of the planar feed antenna acts as a ground plane for the RCA. It is worth pointing out that the ground plane is essential to obtain wideband gain enhancement from this RCA, which was thoroughly studied and discussed in Reference [16]. The complete design parameters of the dielectric PRS are given in Table 2.

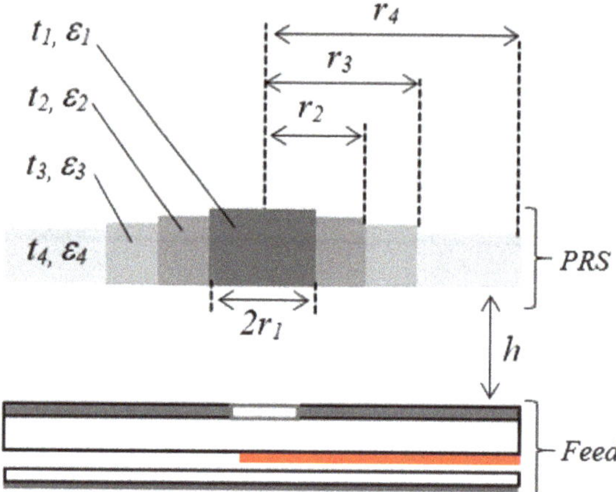

Figure 3. Schematic of RCA constructed by placing the wideband dielectric partially reflecting superstructure (PRS) at a height *h* above the proposed planar strip-line fed U-slot.

Table 2. Design parameters of the wideband dielectric partially reflecting superstructure (PRS) (in mm).

Parameter	r_1	r_2	r_3	r_4	h
	14	20	29.6	39.5	16
Parameter	t_1	t_2	t_3	t_4	h_L
	11.9	11.4	10.16	8.18	1

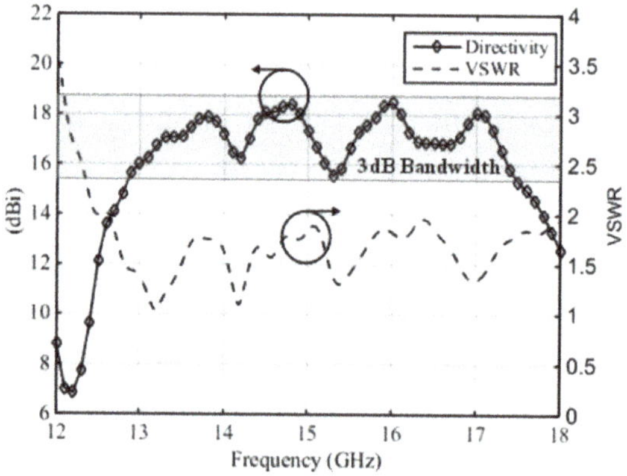

Figure 4. Broadside directivity and input matching of the RCA with planar stripline-fed U-slot.

Figure 4 shows the broadside directivity and VSWR of this RCA, with the planar feed interfaced with the wideband PRS. The peak directivity increased from 8 dBi (feed only, see Figure 2b) to 18.5 dBi in Figure 4, which is a significant 10 dB increase. The radiation pattern of the RCA with U-slot feeding structure at 15 GHz is also given in Figure 2e to show the improvement. The VSWR 2:1 bandwidth spans over 12.4–18 GHz, corresponding to 37% matched bandwidth. The matched bandwidth clearly complements the 3 dB directivity bandwidth, which is imperative to obtain an equivalent 3 dB gain bandwidth and to minimize the mismatch loss. It can be noted in Figure 4 that the curve of broadside directivity versus frequency is not entirely flat, and has some drops at 14.2 GHz, 15.3 GHz, and 16.5 GHz, although these drops are less than 3 dB from the peak value of directivity. A close comparison of Figure 4 with Figure 2b shows that these frequencies approximately correspond to the frequencies where the broadside directivity of the feed antenna drops down, which in turn reflects in the broadside directivity of RCA. As discussed in Section 2, the air gap h_1, and the dimensions of the U-shaped stub at the coupling end of the stripline, were parametrically studied to achieve the best compromise between matching and broadside directivity. These two parameters can readily influence the 3-dB directivity bandwidth, particularly when there a small drops observed in the broadside directivity, seen in Figure 4.

Figure 5 shows the effect of slight variations in the air gap h_1, by varying the value of h_1 from 0.9 mm to 1.2 mm. It can be observed from Figure 5 that larger values of h_1 may lead to a deterioration in the matching of the RCA (see Figure 5a), particularly from 16–16.5 GHz and 17.5–18 GHz, whereas smaller values of h_1 can influence the drop in broadside directivity around 15 GHz, and has the potential of splitting the 3 dB directivity bandwidth into two bands (see Figure 5b). Therefore, setting the value of $h_1 = 1$ mm is a reasonable compromise. Similarly, Figure 6 shows the effect of a U-shaped stub that is added and tuned at the coupling end of the stripline. In the absence of this stub, the broadside directivity of the feed antenna is reduced and, in turn, the broadside directivity of the RCA is decreased. Figure 6 shows that, with the U-shaped stub removed, the directivity bandwidth reduces from 32% (12.8 to 17.65 GHz) to only 21% (14.2 to 17.6 GHz).

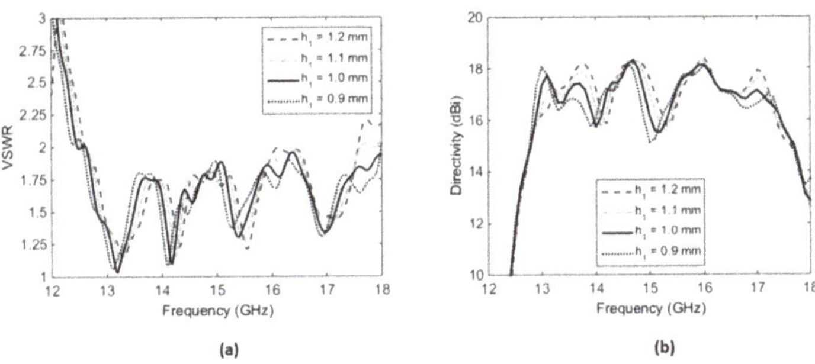

Figure 5. Sensitivity of air-gap h_1 with respect to VSWR and broadside directivity of the RCA, (**a**) VSWR, (**b**) broadside directivity.

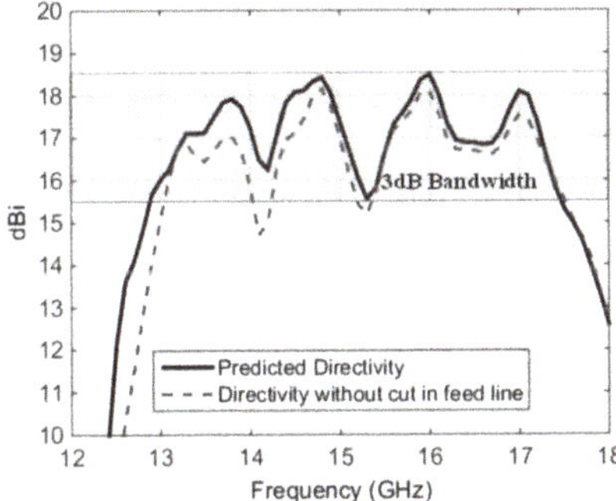

Figure 6. Effect of U-shaped stub in increasing the directivity bandwidth of the antenna.

4. Measurement

A prototype of the proposed RCA interfaced with the planar feed was fabricated and measured in the NSI-700S-50 spherical nearfield anechoic chamber. Three nylon spacers were used to mount the wideband dielectric PRS above the planar feed and a SMA connector was soldered to the stripline. The prototype is shown in Figure 7, along with the stripline assembly taken apart to show the U-shaped stub, and the U-slot etched in the ground plane. The U-slot was etched on one side of the copper cladding on the Rogers RO3003 substrate (Rogers Corporation Inc., Chandler, AZ, USA) ($\varepsilon_r = 3.0$ and tan $\delta = 0.0019$) having a thickness of 3.04 mm. The stripline was laid out on the bottom of the substrate, and excess cladding was etched out, as shown in Figure 7. The U-slot feeding structure and dielectric PRS were constructed using the parameters given in Tables 1 and 2, respectively.

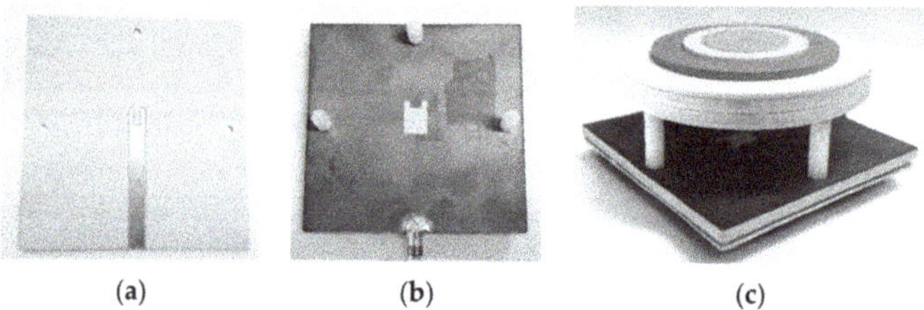

(a) (b) (c)

Figure 7. Photograph of fabricated prototype for measurements. (**a**) Back view, (**b**) top view of substrates of planar wideband feed taken apart and (**c**) the assembled RCA prototype.

The VSWR of the planar feed antenna was measured using an Agilent PNA-X N5242A Vector Network Analyzer (Agilent Technologies Inc.). Figure 8a shows the VSWR for the stand-alone planar feed, as well as with the planar feed interfaced with the dielectric PRS. It can be seen that, for both cases, the VSWR is less than 2 over the frequency range from 12.3 GHz to 17.8 GHz. In case of the standalone feed antenna, the VSWR rises above 2 beyond 17.25 GHz. This is because the U-slot and the U-shaped stub were optimized with the PRS integrated with the feed antenna.

The prototype demonstrated a measured peak broadside directivity and realized gain of 18 dBi and 16.7 dBi, respectively, with a measured 3 dB directivity bandwidth extending from 12.3 GHz to 17.8 GHz, as shown in Figure 8b. The gain was measured using the gain comparison method, using standard-gain horns SGH-75 and SHG-51 as the reference antennas in the respective frequency bands. A slight difference between directivity and gain, approaching a maximum of 1.5 dB at a few data points, can be observed in Figure 8b. This difference is attributed to two facts. Firstly, the gain-comparison method presents an inherent tolerance of ±0.5 dB. Secondly, the VSWR of the antenna is very sensitive to the air gap in the planar feed antenna, as discussed in Section 3. While all due care was taken to maintain the air gap adequately and to keep the bottom and top layers of the feed parallel to each other, the prototype is not perfect. The measured radiation patterns of the presented antenna at three different frequencies within the gain bandwidth, shown in Figure 9, confirm the directive nature of the antenna.

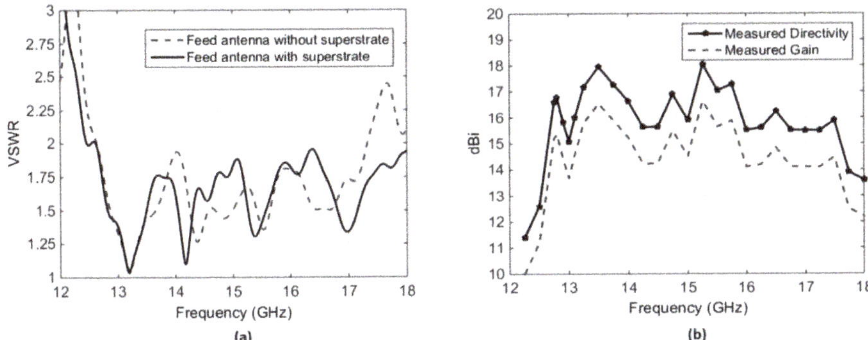

Figure 8. Measured results of the RCA with planar feed antenna, (**a**) VSWR of the antenna with and without PRS, (**b**) broadside directivity and realized gain.

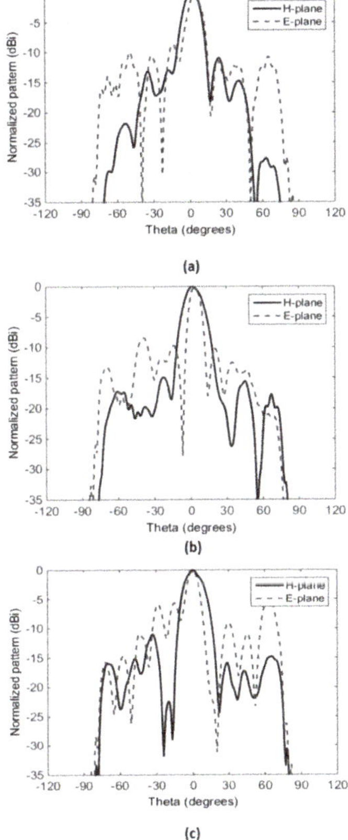

Figure 9. Radiation patterns of the RCA with planar feed measured at Australian Antenna Measurement Facility (AusAMF), (**a**) 12.8 GHz, (**b**) 15.0 GHz, (**c**) 16.0 GHz.

The measured performance comparisons of the proposed antenna with the previously reported RCA is given in Table 3. As shown, the proposed antenna demonstrated significantly large bandwidth compared to that of the RCA fed by a slot-coupled WR-75 waveguide [4] and slot coupled patch antenna [10].

Table 3. Measured performance comparison of the proposed antenna with previously published antennas.

	Feed Type	Peak Directivity (dBi)	Bandwidth (GHz/%)	Footprint (mm × mm)
This work	U-slot	18	12.3–17.8 GHz	80 × 80
[4]	WR-75	20	15%	80 × 80
[10]	Slot Coupled patch	15	13.5–17.5 GHz	45 × 45

5. Conclusions

A simple, planar, wideband feed antenna has been proposed, which can be used to feed wideband RCAs. It has the potential to be fabricated through the standard printed-circuit board (PCB) manufacturing process, and can easily be interfaced with any type of wideband PRS to form a stand-along RCA. The approach is a simple alternative to the expensive and bulky waveguide feeding methods used in conventional wideband RCAs. A maximum antenna gain of 16.7 dBi has been obtained with a directivity bandwidth covering nearly the entire Ku-band. Measured results of the fabricated porotype are presented and demonstrate a wideband performance with reasonably high gain and excellent matching.

Author Contributions: Methodology, A.A.B.; conceptualization and software, A.A.B. and R.M.H.; data curation, M.A.; writing—original draft preparation, A.A.B.; writing—review & editing, L.M.; supervision, R.M.H.; project administration, K.P.E.

Funding: This work was supported by the Australian Research Council (ARC) (Grant Numbers: DP190103352 and DP 150103242).

Acknowledgments: The authors would like to thank Rogers Inc., for providing the dielectric materials used in the antenna prototype for experiments.

Conflicts of Interest: The authors declare no conflict of interest.

References

1. Baba, A.A.; Hashmi, R.M.; Esselle, K.P.; Weily, A.R. Compact High-Gain Antenna with Simple All-Dielectric Partially Reflecting Surface. *IEEE Trans. Antennas Propag.* **2018**, *66*, 4343–4348. [CrossRef]
2. Weily, A.R.; Esselle, K.P.; Sanders, B.C.; Bird, T.S. High-gain 1D EBG resonator antenna. *Microw. Opt. Technol. Lett.* **2005**, *47*, 107–114. [CrossRef]
3. Ge, Y.; Esselle, K.; Bird, T. The use of simple thin partially reflective surfaces with positive reflection phase gradients to design wideband, low-profile EBG resonator antennas. *IEEE Trans. Antennas Propag.* **2012**, *60*, 743–750. [CrossRef]
4. Konstantinidis, K.; Feresidis, A.; Hall, P. Multilayer partially reflective surfaces for broadband fabry-perot cavity antennas. *IEEE Trans. Antennas Propag.* **2014**, *62*, 3474–3481. [CrossRef]
5. Al-Tarifi, M.A.; Anagnostou, D.E.; Amert, A.K.; Whites, K.W. Two-cavity model for creating two high-directivity bands of the resonant cavity antenna with flexible and dynamic control. In Proceedings of the 2013 IEEE Antennas and Propagation Society International Symposium (APSURSI), Orlando, FL, USA, 7–13 July 2013; pp. 1512–1513.
6. Weily, A.; Horvath, L.; Esselle, K.; Sanders, B.; Bird, T. A planar resonator antenna based on a woodpile EBG material. *IEEE Trans. Antennas Propag.* **2005**, *53*, 216–223. [CrossRef]
7. Meng, F.; Sharma, S.K. A Dual-Band High-Gain Resonant Cavity Antenna with a Single Layer Superstrate. *IEEE Trans. Antennas Propag.* **2015**, *63*, 2320–2325. [CrossRef]
8. Baba, A.A.; Hashmi, R.M.; Esselle, K.P. Wideband gain enhancement of slot antenna using superstructure with optimised axial permittivity variation. *Electron. Lett.* **2016**, *52*, 266–268. [CrossRef]
9. Ge, Y.; Esselle, K.P. A method to design dual-band, high directivity EBG resonator antennas using single-resonant, single layer partially reflective surfaces. *Prog. Electromagn. Res. C* **2010**, *13*, 245–257. [CrossRef]
10. Wang, N.; Li, J.; Wei, G.; Talbi, L.; Zeng, Q.; Xu, J. Wideband Fabry-Perot resonator antenna with two layers of dielectric superstrates. *IEEE Antennas Wirel. Propag. Lett.* **2015**, *14*, 229–232. [CrossRef]

11. Zhao, T.; Jackson, D.; Williams, J.; Yang, H. Radiation characteristics of a 2D periodic leaky-wave antenna using metal patches or slots. *IEEE Antennas Propag. Soc. Int. Symp.* **2001**, *3*, 260–263.
12. Hashmi, R.M.; Zeb, B.A.; Esselle, K.P. Wideband high-gain EBG resonator antennas with small footprints and all-dielectric superstructures. *IEEE Trans. Antennas Propag.* **2014**, *62*, 2970–2977. [CrossRef]
13. Al-Tarifi, M.A.; Anagnostou, D.E.; Amert, A.K.; Whites, K.W. The puck antenna: A compact design with wideband, high-gain operation. *IEEE Trans. Antennas Propag.* **2015**, *63*, 1868–1873. [CrossRef]
14. Zeb, B.A.; Hashmi, R.M.; Esselle, K.P. Wideband gain enhancement of a slot antenna using one unprinted dielectric superstrate. *IET Electron. Lett.* **2015**, *51*, 1146–1148. [CrossRef]
15. Hashmi, R.M.; Esselle, K.P. A class of extremely wideband resonant cavity antennas with large directivity-bandwidth products. *IEEE Trans. Antennas Propag.* **2016**, *64*, 830–835. [CrossRef]
16. Baba, A.A.; Hashmi, R.M.; Esselle, K.P. Achieving a large gain bandwidth product from a compact antenna. *IEEE Trans. Antennas Propag.* **2017**, *65*, 3437–3446. [CrossRef]
17. Hashmi, R.M.; Esselle, K.P. A wideband EBG resonator antenna with an extremely small footprint area. *Microw. Opt. Technol. Lett.* **2015**, *57*, 1531–1535. [CrossRef]
18. Wu, F.; Luk, K.M. Wideband high-gain open resonator antenna using a spherically modified, second-order cavity. *IEEE Trans. Antennas Propag.* **2017**, *65*, 2112–2116. [CrossRef]
19. Ge, Y.; Sun, Z.; Chen, Z.; Chen, Y.Y. A high-gain wideband low profile fabry-perot resonator antenna with a conical short horn. *IEEE Antennas Wirel. Propag. Lett.* **2016**, *15*, 1889–1892. [CrossRef]

Article

A 135-190 GHz Broadband Self-Biased Frequency Doubler using Four-Anode Schottky Diodes

Chengkai Wu, Yong Zhang *, Jianhang Cui, Yukun Li, Yuehang Xu and Ruimin Xu

School of Electronic Science and Engineering, University of Electronic Science and Technology of China, Chengdu 611731, China; chengkaiwu@std.uestc.edu.cn (C.W.); cuijianhang@std.uestc.edu.cn (J.C.); 18215605919@163.com (Y.L.); yuehangxu@uestc.edu.cn (Y.X.); rmxu@uestc.edu.cn (R.X.)

* Correspondence: yongzhang@uestc.edu.cn

Received: 22 March 2019; Accepted: 22 April 2019; Published: 25 April 2019

Abstract: This paper describes the design and demonstration of a 135–190 GHz self-biased broadband frequency doubler based on planar Schottky diodes. Unlike traditional bias schemes, the diodes are biased in resistive mode by a self-bias resistor; thus, no additional bias voltage is needed for the doubler. The Schottky diodes in this verification are micron-scaled devices with an anode area of 6.6 μm^2 and an epitaxial layer thickness of 0.26 μm. For accurate design of the doubler, the 3D-EM model of the Schottky diode is built up to extract the parasitic parameters induced by the diode package when frequency rises up to the terahertz band. In order to implement broadband working, input waveguide steps, output suspended microstrip steps, and output probe with bias filter are all used as matching elements for impedance matching. Measured results show that the doubler exhibits a 3 dB bandwidth of 34% from 135 GHz to 190 GHz, with a conversion efficiency of above 4% when supplied with 100 mW of input power. A 17.8 mW peak output power with a 10.2% efficiency was measured at 166 GHz when the input power was 174 mW. The measured results agree well with the simulated results, which indicates that the self-bias scheme for Schottky diode-based frequency multipliers is feasible and effective.

Keywords: frequency doubler; broadband matching; Schottky diodes; self-bias resistor; conversion loss; three-dimensional electromagnetic (3D-EM) model; millimeter wave; terahertz

1. Introduction

As the last spectrum resource that has not been fully exploited and utilized, the terahertz wave has many potential applications, such as high-speed communication, biomedicine, radio astronomy, and safety imaging [1,2]. The tremendous interest in the terahertz spectrum has prompted researchers to develop a series of components working in the terahertz band, such as amplifiers [3,4], frequency multipliers [5], mixers [6], antennas [7–10] and so on, to construct practical terahertz front-end systems [11–13]. As a key component of terahertz heterodyne systems, frequency multipliers with high enough power are required as front-stage drivers of terahertz frequency multiplier chains [14] or local oscillators (LO) of terahertz mixers. Theoretically, any non-linear device can be used to achieve the function of frequency multiplication by extracting the required harmonic components. Frequency multiplication technology based on planar Schottky diodes is one of the most attractive device technologies to produce millimeter wave/terahertz waves due to its high reliability, high spectral quality, low cost, and operation at room temperature [15,16]. Moreover, great progress has been made in the fabrication and modeling technology of Schottky diodes [14,17,18] as well as in the understanding and characterization of inner physical mechanisms [19–21].

Frequency multipliers mainly focus on the performance of efficiency, output power and bandwidth, and various state-of-the-art multipliers [22–32] based on planar Schottky barrier diodes that have been designed and demonstrated in recent years. Though frequency multipliers like in [22,23] eliminate possible assembly errors and feature better consistency with microwave monolithic integrated circuit (MMIC) technology, the discrete diodes [24–32] are also applied for its low cost and reliable alignment process when the frequency is below 400 GHz [27]. From the perspective of bias approaches, all these frequency multipliers mostly work at zero bias or external reverse bias. As we know, the nonlinear effect of diodes generates not only harmonic components, but also a DC component. However, there are few reports in the literature on a diode's own DC component being used to bias itself in the terahertz band. Moreover, previous experience tells us that frequency multipliers can also work in a self-bias state, which inspired us to design a frequency multiplier that works directly under self-bias conditions.

To demonstrate the self-bias scheme, a 135–190 GHz frequency doubler based on Schottky diodes was designed, in which a resistor parallel to the ground in the bias circuit is utilized to withstand the DC component generated by the diodes to form a self-bias loop. An accurate three-dimensional electromagnetic (3-D EM) model of the diodes with a self-bias resistor was established using the finite element method (FEM), and the influence of the self-bias resistor on the performance of the frequency doubler was also investigated through a harmonic balance simulation. In order to achieve broadband impedance-matching, the optimum embedded impedance of the diodes was extracted using harmonic load-pull. Based on that, the input and output matching circuits were designed. Finally, the doubler, which worked under self-bias condition and exhibited the merits of broad band, high efficiency and high output power, was fabricated, assembled and measured.

2. Materials and Methods

2.1. Diode Modeling

Frequency multipliers based on Schottky diodes use the nonlinear effect of the Schottky barrier junction to produce the required harmonics. In our design, the varactor diode chip, which was produced by Teratech (Oxford, UK) had four anodes arranged in anti-series, as depicted in Figure 1a, and the overall chip dimensions were 420 μm × 80 μm × 50 μm. The diodes used an epitaxial layer with a doping density of 2×10^{17} cm^{-3}, a diode anode size of 6.6 μm^2 and an epitaxial thickness of 0.26 μm, resulting in a calculated zero-bias capacitance of 9.8 fF, a series resistance of 4.1 Ω, and a corresponding cut-off frequency 2.2 THz. These diodes were fabricated on GaAs substrate and had a gold air-bridged structure designed to minimize the parasitic capacitance.

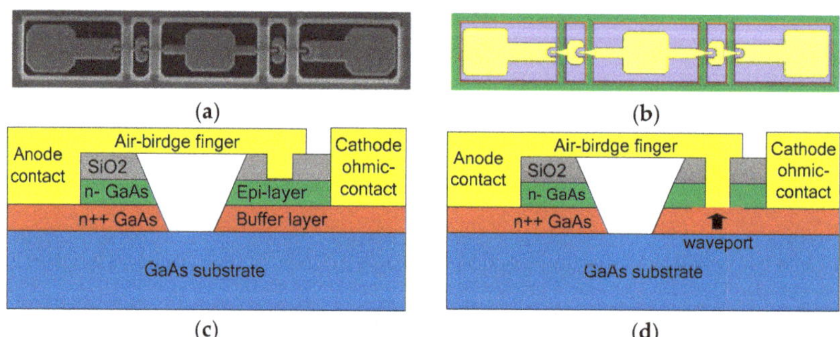

Figure 1. Diode structure and internal port (**a**) SEM Image of the Teratech GaAs anti-series air-bridged Schottky diodes; (**b**) 3-D structure model of the diodes; (**c**) 3-D cross-sectional view of the diodes; (**d**) internal port setting.

Figure 1c shows the 3-dimentional (3-D) cross-sectional view of the diodes, including the air-bridge anode finger, the SiO_2 passivation layer, the light doped n-GaAs epitaxy layer, the heavily doped n++ GaAs buffer layer, and the undoped GaAs substrate from top to bottom. Some approximation processing was needed because the field simulation software could not simulate the doping of the semiconductor. Some detailed information on the thickness and material of various layers are given in

Table 1. According to the electromagnetic characteristics of the materials, a 3-D structure model of the diodes was established with appropriate material and size settings, as shown in Figure 1b. In addition, some key SPICE parameters of the diodes are given in Table 2 for the characterization of the intrinsic part of the Schottky junction.

Table 1. Layer Structure Parameters of the diodes used in this paper.

Layers of the Diodes	Thickness (μm)	Doping Concentration (cm^{-3})	Material	Relative Dielectric Constant
Semi-Insulating Substrate	50	-	GaAs	12.9
Buffer Layer	5	5×10^{18}	Pec	1
Epi-Layer	0.26	2×10^{17}	GaAs	12.9
Oxide Layer	0.5	-	SiO_2	4
Ohmic Contact Layer	0.76	-	Gold	1
Anode and Cathode	1	-	Gold	1

Table 2. Some key SPICE parameters of the diodes.

Parameters	Value
Reverse Saturation Current, I_s	9.39×10^{-15} A
Series Resistance, R_s	4.1 Ω
Ideal Factor, N	1.12
Zero-biased Junction Capacitance, C_{j0}	9.8 fF
Junction Potential, V_j	0.85 V
Reverse Breakdown Current, I_{br}	1 μA
Energy Gap, E_g	1.43 eV

When frequency rose up to the submillimeter wave band, the complicated parasitic effects induced by diode package could not be ignored and the cavity effect of complex structures also had to be taken into consideration. The accurate three-dimensional electromagnetic model (3-D EM) of the diodes in a balanced structure was built up in a High Frequency Structure Simulator (HFSS) (ANSYS, Canonsburg, PA, USA, 2015) [33] to extract the parasitic parameters and simulate the electromagnetic environment where the diode chip was located, as shown in Figure 2. Since HFSS does not support enclosed wave ports, coaxial probe technique [17] was adopted in the process of modelling and the anode probe penetrated through the epitaxial layer to the buffer layer to set internal wave ports at the interface, as depicted in Figure 1d.

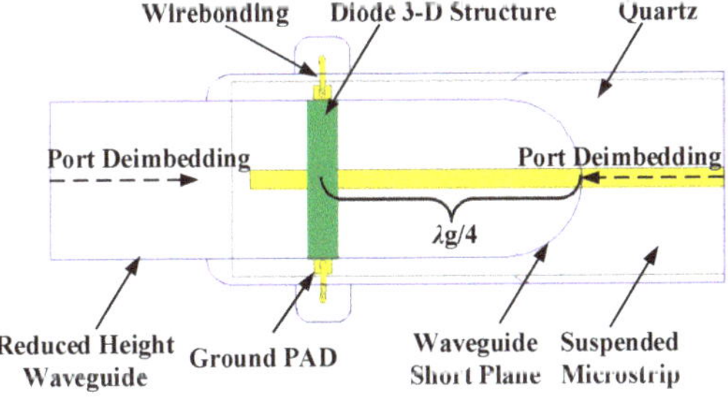

Figure 2. The 3-D EM model of the diodes in balanced structure in HFSS.

The equivalent circuit model of the diodes was constructed with the parasitic parameters derived from the three-dimensional electromagnetic model of SNP items and the intrinsic SPICE parameters of the Schottky barrier, as shown in Figure 3. In the equivalent circuit model, a resistor was introduced to form a self-bias loop. The method of harmonic load-pull was then executed using Agilent's Advanced

Design System (ADS) (Agilent, Palo Alto, CA, USA, 2013) [34] on the equivalent circuit model to determine the optimum embedding impedance, which was 227.3-j24.2 Ω for the input at 80 GHz and 34.6-j26.8 Ω for the output at 160 GHz, respectively. In carrying out the load-pull procedure, the self-bias resistor was replaced by a DC voltage source for convenience. The optimum bias voltage was found to be −2 V for 100 mW input power, with a conversion efficiency of 20%.

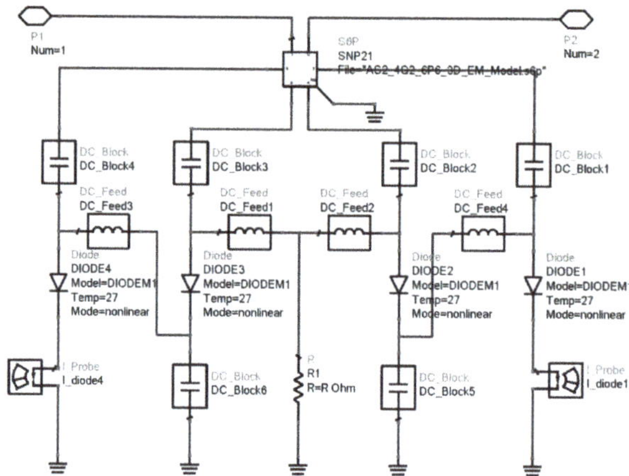

Figure 3. The equivalent circuit model of the diodes with self-bias resistor in ADS.

2.2. Doubler Design

In a balanced structure, frequency multipliers have only an even-order or odd-order harmonics output; thus, there is no need for additional filters which tend to add circuit loss and degrade the multiplication bandwidth. In the balanced frequency doubler proposed originally by Erikson [35,36], the Schottky diode in anti-series configuration was placed at the junction where the balanced transmission line converted to unbalanced. In this structure, the incident fundamental wave was coupled onto the anti-series Schottky diode pair in a TE10 mode of the input waveguide and prevented from propagating into the output circuit by the backshort, which was achieved via narrowing down the input waveguide width. The even-order harmonics were then free to propagate down the output circuit in a TEM mode, while the odd-order harmonics were suppressed. The isolation between the input and output circuit in the balanced doubler was naturally realized by the electromagnetic mode orthogonality.d

According to these parameters, a 135–190 GHz broadband self-biased frequency doubler was proposed, as shown in Figure 4. The doubler was composed of input waveguide steps, output suspended microstrip steps, an output probe, bias filter, Schottky diode model, and a self-bias resistor. The input waveguide was WR10 and the output waveguide was WR5.1. The fundamental wave energy coupled to the diodes generated not only an even harmonic output, but also a DC component. The DC was loaded on the self-bias resistor, which in turn provided bias for the diodes via the bias filter. For the convenience of subsequent measurements, this resistor was placed outside the cavity. In practical applications, it could be easily integrated into the cavity. The external self-bias resistor was connected to the internal bias network through a piercing capacitor fixed on the cavity and a bias line made of Rogers 5880 substrate with a thickness of 0.127 mm. In order to prevent second harmonic leakage, a two-order Common Mode Resonant Cell (CMRC) structure was used in the bias circuit, which formed two transmission zeroes in the output frequency band, and thus achieved good isolation between the output frequency and DC bias. The output circuit adopted a suspended microstrip rather than a microstrip for two reasons. On the one hand, the suspended microstrip can reduce the loss of the matching circuit, considering that most of its fields are distributed in the air. On the other hand, compared with the microstrip, the suspended microstrip can improve the cut-off frequency of the higher-modes and guarantee that the second harmonic propagates in a single mode. The output suspended microstrips, output probe, and bias filter were integrated on a 50 μm thick quartz substrate. In addition, two steps with a depth of 70 μm were added on both sides of the reduced height input waveguide underneath the diodes in order to support the quartz substrate, while the other side of quartz substrate was supported by the ground of the CMRC microstrip.

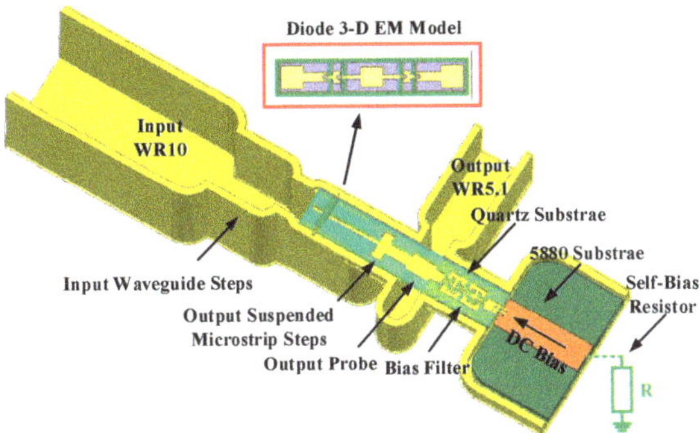

Figure 4. The configuration of the 135-190 GHz balanced frequency doubler under self-bias.

In order to investigate the impedance characteristics of the diodes in a wide frequency band, load-pull analysis was performed on the equivalent circuit model, as shown in Figure 3, to give12% efficiency at different input frequencies under the optimum bias voltage. The results of load-pull at different frequencies were sketched on the same Smith chart shown in Figure 5, which shows that equal efficiency circles rotate counter-clockwise on the Smith chart as the frequency increases from 75 to 90 GHz and an intersection region is formed. The intersection of equal efficiency circles at different input frequencies is the broadband output matching region, where the efficiency exceeds 12% for the 75–90 GHz input frequency range, provided that the output load impedance falls into the region. Analogously, the broadband input matching region could also be obtained by a similar source-pull operation. The existence of the broadband matching region indicated that the varactor was biased in resistive mode and that broadband input matching and output matching were theoretically achievable.

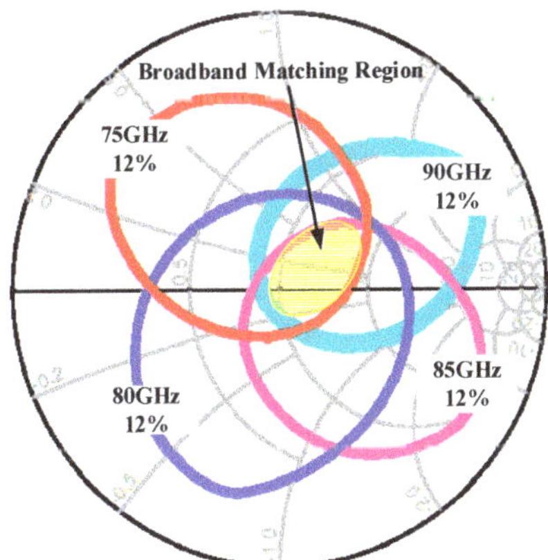

Figure 5. Output load-pull for a 12% efficiency at different input frequencies.

Figure 6 shows the design of input and output matching. The input matching circuit included two-stage reduced height waveguides, while the height of the second stage waveguide was fully reduced to suppress the unwanted TM11 mode. The output matching circuit included not only suspended microstrip steps, but also the output waveguide E-plane probe, which took part in output matching while realizing the mode conversion (from TEM to TE10). The S-parameters of the waveguide steps, suspended microstrip steps and E-plane waveguide probe were analyzed and extracted in SNP files using HFSS to characterize the discontinuity effects. Moreover, the suspended microstrip model was also constructed by extracting feature parameters in the field simulation, such as characteristic impedance, attenuation constant and effective permittivity. Based on the above work, the whole harmonic balance simulation circuit of the frequency doubler was established. L1–L7 in Figure 6 were the main matching variables in the whole circuit, which were adjusted and optimized iteratively by HFSS in combination with ADS, so that Zin on the reference plane 1-1′ and Zout on the reference plane 2-2′ fell into the broadband matching region in the required bandwidth. Thus, broadband input and output matching was implemented.

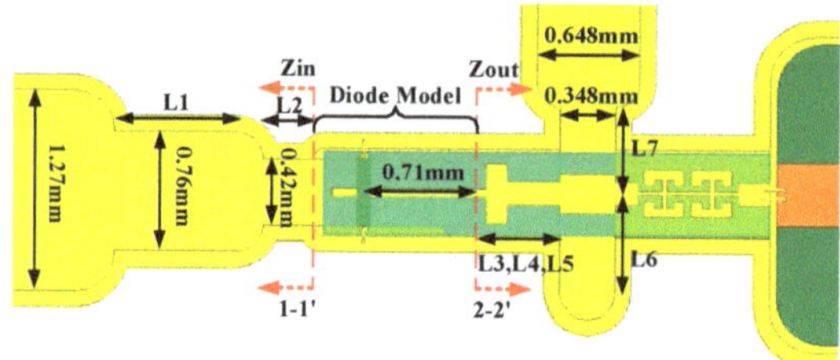

Figure 6. Design of input and output matching (L1 = 767 μm, L2 = 318 μm, L3 = 58 μm, L4 = 138 μm, L5 = 330 μm, L6 = 622 μm, L7 = 547 μm).

The DC voltage source was then replaced with the self-bias resistor. Simulation results showed that conversion efficiency ascended with the increase of the self-bias resistor from 10 to 100 Ω, while it declined with the increase of self-bias resistance from 100 to 400 Ω, as shown in Figure 7a,b, respectively. The efficiency curve agreed well with the simulation results under the optimum bias voltage, especially when the self-bias resistor was 100 Ω. Therefore, it could be speculated that the optimum self-bias resistor was 100 Ω. Figure 8 shows the simulated results of input return loss, conversion loss and output power with 100 mW input power under a 100 Ω self-bias resistor.

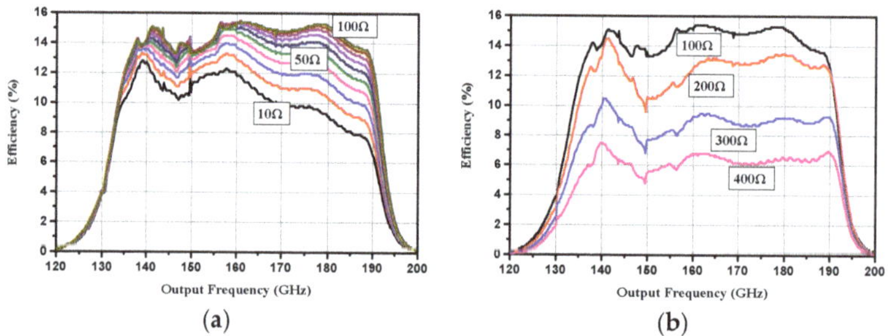

Figure 7. Simulated efficiency under different self-bias resistors (**a**) from 10 to 100 Ω; (**b**) from 100 to 400 Ω.

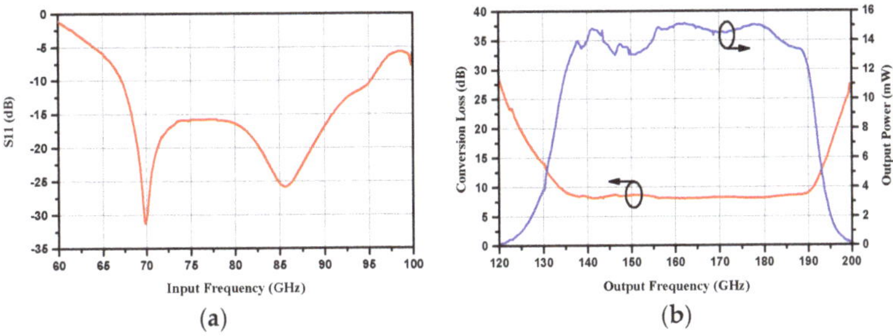

Figure 8. Simulated input return loss, conversion loss and output power with 100 mW input power under a 100 Ω self-bias resistor.

2.3. Fabrication and Measurement

The manufacturing of the whole frequency doubler required numerically controlled metal machining (CNC) to process the cavity structure, thin film circuit technology to process the quartz substrates, and printed circuit board (PCB) technology to make the 5880 bias lines. After finishing the design of the doubler, the fabricating maps of the planar circuit and cavity structure were drawn by AutoCAD (2016) and Solidworks (2016). The whole frequency doubler circuit, including the bias filter, was manufactured on a 50 μm thick quartz substrate, with dimensions of 2.8 mm × 0.52 mm, embedded in the split waveguide block made of brass plated with 2 μm thickness of gold layer. The assembly procedures of the doubler mainly included cavity cleaning, substrate adhering, diode adhering, circuit sintering, capacitor installation, and wire bonding. In the assembly, substrate alignment was very important, otherwise the performance would be greatly deteriorated by substrate offset. Besides, it should be noted that the baking temperature did not exceed 200 °C to prevent the substrate from bending due to excessive temperature. The internal structure of the assembled frequency doubler is shown in Figure 9. The diode chip was mounted onto the quartz substrate circuit with the cathode pads connected to the split waveguide block, using the conductive silver adhesives to realize the DC and RF grounding, and to provide a good heat dissipation channel. The bias filter on the quartz substrate was connected to a bias line made of Rogers 5880 substrate, using a wire bond to provide a bias path.

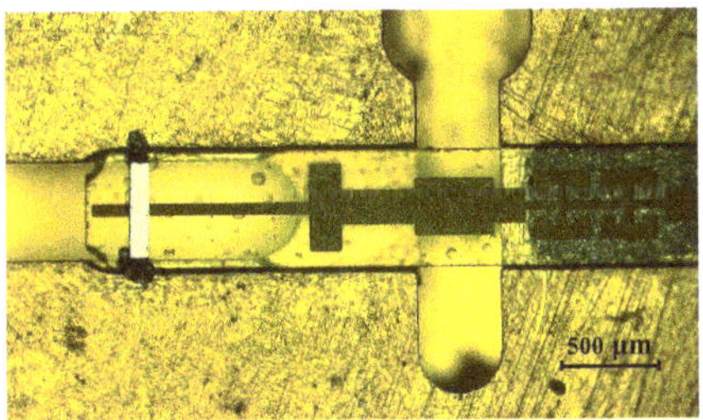

Figure 9. The image of the 135–190 GHz doubler imbedded in the split-waveguide block.

For the purpose of verifying the proposed the self-bias scheme for Schottky diodes, two prototypes of the frequency doubler were manufactured for comparative measurements. The test platform of the doublers is shown in Figure 10. In order to measure the frequency doublers by turn, the test frequency band was divided into three parts: 130–160 GHz, 160–180 GHz, and 180–200 GHz bands. Driving power was provided by an Agilent E8257D PSG Analog Signal Generator (Santa Clara, CA, USA), followed by different MMIC frequency multiplication and amplification modules. The input power was calibrated in advance by the VDI Erickson PM4 millimeter/submillimeter power meter (Virginia Diodes, Inc., Charlottesville, VA, USA). The output power of the doublers was measured using the same power meter, together with a 2.5 cm long WR5.1 to WR10 waveguide transition to minimize the mismatch between the doubler and meter (the measured output power was not corrected for the transition loss). An external potentiometer was used as a self-bias resistor, which could be easily adjusted to find the optimum self-bias resistance.

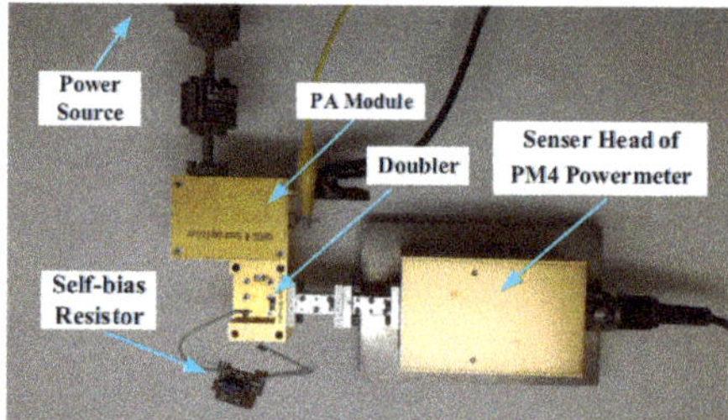

Figure 10. Test platform of the doubler.

3. Results and Discussion

The measured conversion efficiency and output power are shown in Figure 11. A conversion efficiency of around 8% and a mean output power of 8 mW from 140 to 180 GHz were measured under a 100 Ω self-bias resistor with 100 mW input power. The doubler exhibited a 3 dB conversion loss bandwidth of 34% from 135 GHz to 190 GHz. The self-bias voltage generated by the diodes under 100 Ω self-bias resistor was measured close to −2 V when the input power was 100 mW, which was in accordance with the simulation results. As the input power increased, the self-bias voltage decreased slightly.

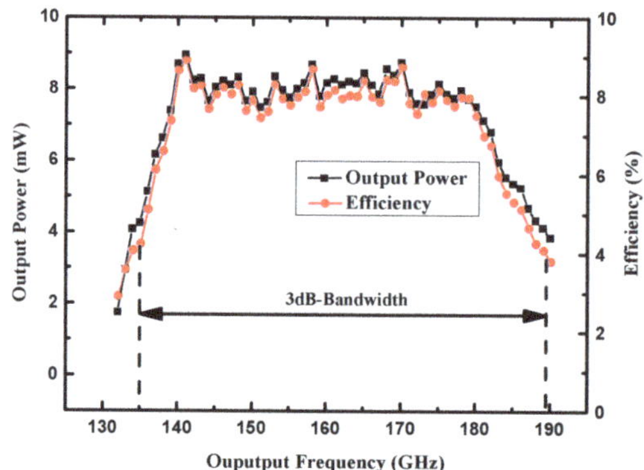

Figure 11. Measured efficiency and output power of the doubler under a self-bias resistor of 100 Ω.

When the maximum input power was provided, the frequency doubler was tested under different self-bias conditions, as shown in Figure 12. The output power increased with the increase of the self-resistor from 10 to 100 Ω and it decreased distinctly with the increase of the self-bias resistor from 100 to 400 Ω, with 160 mW input power at 160 GHz. A similar phenomena occurred with a peak output power of 17.8 mW and a peak efficiency of 10.2% at 166 GHz when the input power was 174 mW, except that the optimum self-bias resistor was 200 Ω. This was a slight deviation from the simulation results shown in Figure 8, and was probably caused by a power saturation effect, because each anode shared more than 40 mW input power. The measured output power and efficiency as a function of input power under the 50 Ω self-bias resistor are shown in Figure 13. The output power increased with the input power and the efficiency was stable at 8%. If greater power was provided, the doubler would have a higher output power level at both ends of the working frequency band. It should be noted that the curves in Figures 12 and 13 represent the average values of the measured results of the two prototypes, and the error bars represent the standard deviations of the measured results for each prototype.

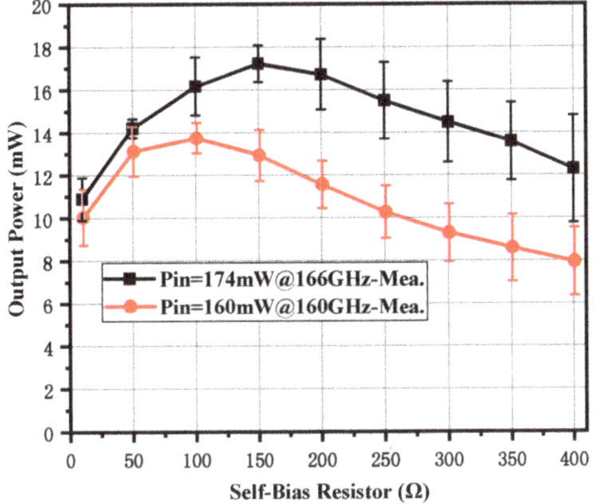

Figure 12. Measured output power of the doubler under different self-bias resistors with the maximum input power.

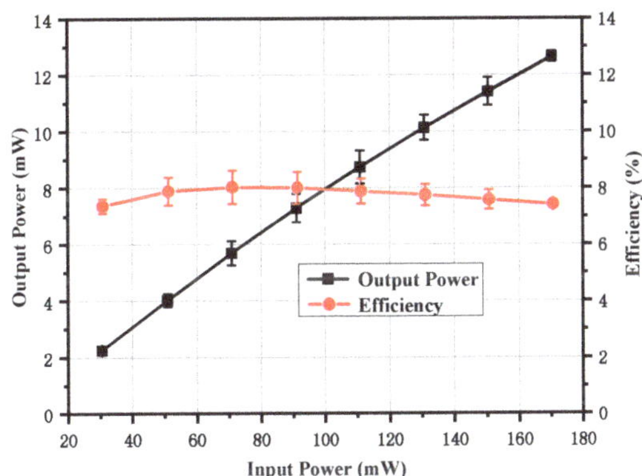

Figure 13. Measured output power and efficiency versus input power at 166 GHz under 50 Ω self bias resistor.

A comprehensive comparison of the designed frequency doubler with the multipliers reported in the literature is given in Table 3. Compared with the tripler [24] and the doublers [25–27] working under external bias, the proposed doubler had a larger fractional bandwidth. Although a record full waveguide bandwidth of 44% was obtained by the doubler working under zero bias in ref. [29], its output power and power capacity were smaller than this work due to the fewer number of anodes. The doubler working under zero bias in Ref. [22] used the largest number of anodes in the whole table and achieved a similar efficiency as shown in this work. However, higher output power and larger fractional bandwidth were obtained by the proposed doubler using only four anodes. From the above comparison, it can be concluded that the proposed self-bias scheme for frequency multipliers achieved a compromise in bandwidth and efficiency. For the designed doubler itself, the combination of broadband and high output power made it versatile.

Table 3. Comprehensive comparison of reported frequency multipliers.

Ref.	Frequency (GHz)	Number of Anodes	Pin (mW)	Pout (mW)	Effi. (%)	FBW (%)	Bias Way
[22]	177–202	6 × 2	20–120	1–12	4.3–10.2	22	Zero Bias
[24]	200–235	6	20–60	1.5–15	~16	24	External Bias
[25]	150–170	6	10–200	1–35	5.1–23.5	10	External Bias
[26]	190–198	6	100–260	3.7–20	1.4–8	4	External Bias
[27]	180–200	6	50–95	4–13	4–10	10.5	External Bias
[29]	140–220	2	20–32	1.8–3.7	8.7–12.7	44	Zero Bias
This Work	135–190	4	30–174	2.1–17.8	4–10.2	34	Self Bias

4. Conclusions

In this article, a 135–190 GHz broadband self-biased frequency doubler based on planar Schottky diodes was demonstrated. A resistor was utilized to withstand the DC component generated by the diodes to form a self-bias scheme, which contributed to simplifying the bias scheme of the frequency multipliers. Meanwhile, input waveguide steps, output suspended microstrip steps and an output probe with bias filter were all used as matching elements for broadband impedance matching. The validity and accuracy of the diode model were verified by the agreement between the simulation and measurements. The frequency doubler presented in this work has the merits of broad band, high efficiency, high output power and self-bias working, which makes it very attractive for practical broadband terahertz applications.

Author Contributions: Conceptualization, C.W. and Y.Z.; Methodology, C.W.; Software, J.C.; Validation, C.W. and Y.L.; Investigation, C.W.; Writing—original draft preparation, C.W.; Writing—review and editing, Y.X. and R.X.; Funding acquisition, Y.Z.

Funding: This research was funded by National Natural Science Foundation of China, grant number 61871072 and Sichuan Science and Technology Plan Project (2017JY0277).

Conflicts of Interest: The authors declare no conflict of interest.

References

1. Siegel, P.H. Terahertz technology. *IEEE Trans. Microw. Theory Tech.* **2002**, *50*, 910–928. [CrossRef]
2. Song, H.; Nagatsuma, T. Present and Future of Terahertz Communications. *IEEE Trans. Terahertz Sci. Technol.* **2011**, *1*, 256–263. [CrossRef]
3. Paoloni, C.; Di Carlo, A.; Brunetti, F.; Mineo, M.; Ulisse, G.; Durand, A.; Krozer, V.; Kotiranta, M.; Fiorello, A.M.; Dispenza, M.; et al. Design and Fabrication of a 1 THz Backward Wave Amplifier. *Terahertz Sci. Technol.* **2011**, *4*, 149–163.
4. Samoska, L.A. An Overview of Solid-State Integrated Circuit Amplifiers in the Submillimeter-Wave and THz Regime. *IEEE Trans. Terahertz Sci. Technol.* **2011**, *1*, 9–24. [CrossRef]
5. Chattopadhyay, G. Technology, Capabilities, and Performance of Low Power Terahertz Sources. *IEEE Trans. Terahertz Sci. Technol.* **2011**, *1*, 33–53. [CrossRef]
6. Maestrojuan, I.; Ederra, I.; Gonzalo, R. Fourth-Harmonic Schottky Diode Mixer Development at Sub-Millimeter Frequencies. *IEEE Trans. Terahertz Sci. Technol.* **2015**, *5*, 518–520. [CrossRef]
7. Sadeghzadeh, R.A.; Alibakhshi-Kenari, M.; Naser-Moghadasi, M. UWB antenna based on SCRLH-TLs for portable wireless devices. *Microw. Opt. Technol. Lett.* **2016**, *58*, 69–71. [CrossRef]
8. Alibakhshikenari, M.; Virdee, B.S.; Khalily, M.; Shukla, P.; See, C.H.; Abd-Alhameed, R.A.; Falcone, F.; Limiti, E. Beam-scanning leaky-wave antenna based on CRLH-metamaterial for millimeter-wave applications. *IET Microw. Antennas Propag.* **2019**. [CrossRef]
9. Liu, Y.; Lu, H.; Wu, Y.; Cui, M.; Li, B.; Zhao, P.; Lv, X. Millimeterwave and Terahertz Waveguide-Fed Circularly Polarized Antipodal Curvedly Tapered Slot Antennas. *IEEE Trans. Antennas Propag.* **2016**, *64*, 1607–1614. [CrossRef]
10. Deng, X.; Li, Y.; Liu, C.; Wu, W.; Xiong, Y. 340 GHz On-Chip 3-D Antenna With 10 dBi Gain and 80% Radiation Efficiency. *IEEE Trans. Terahertz Sci. Technol.* **2015**, *5*, 619–627. [CrossRef]

11. Hirata, A.; Kosugi, T.; Takahashi, H.; Takeuchi, J.; Togo, H.; Yaita, M.; Kukutsu, N.; Aihara, K.; Murata, K.; Sato, Y.; et al. 120-GHz-Band Wireless Link Technologies for Outdoor 10-Gbit/s Data Transmission. *IEEE Trans. Microw. Theory Tech.* **2012**, *60*, 881–895. [CrossRef]
12. Ducournau, G.; Szriftgiser, P.; Beck, A.; Bacquet, D.; Pavanello, F.; Peytavit, E.; Zaknoune, M.; Akalin, T.; Lampin, J.F. Ultrawide-Bandwidth Single-Channel 0.4-THz Wireless Link Combining Broadband Quasi-Optic Photomixer and Coherent Detection. *IEEE Trans. Terahertz Sci. Technol.* **2014**, *4*, 328–337. [CrossRef]
13. Li, C.; Ko, C.; Kuo, M.; Chang, D. A 340-GHz Heterodyne Receiver Front End in 40-nm CMOS for THz Biomedical Imaging Applications. *IEEE Trans. Terahertz Sci. Technol.* **2016**, *6*, 625–636. [CrossRef]
14. Chattopadhyay, G.; Schlecht, E.; Ward, J.S.; Gill, J.J.; Javadi, H.H.; Maiwald, F.; Mehdi, I. An all-solid-state broad-band frequency multiplier chain at 1500 GHz. *IEEE Trans. Microw. Theory Tech.* **2004**, *52*, 1538–1547. [CrossRef]
15. Crowe, T.W.; Bishop, W.L.; Porterfield, D.W.; Hesler, J.L.; Weikle, R.M. Opening the terahertz window with integrated diode circuits. *IEEE J. Solid State Circuits* **2005**, *40*, 2104–2110. [CrossRef]
16. Mehdi, I.; Siles, J.V.; Lee, C.; Schlecht, E. THz Diode Technology: Status, Prospects, and Applications. *Proc. IEEE* **2017**, *105*, 990–1007. [CrossRef]
17. Hesler, J.L.; Hall, W.R.; Crowe, T.W.; Weikle, R.M.; Deaver, B.S.; Bradley, R.F.; Pan, S.K. Fixed-tuned submillimeter wavelength waveguide mixers using planar schottky-barrier diodes. *IEEE Trans. Microw. Theory Tech.* **1997**, *45*, 653–658. [CrossRef]
18. Siles, J.V.; Grajal, J. Physics-Based Design and Optimization of Schottky Diode Frequency Multipliers for Terahertz Applications. *IEEE Trans. Microw. Theory Tech.* **2010**, *58*, 1933–1942. [CrossRef]
19. Xu, H.; Schoenthal, G.S.; Liu, L.; Xiao, Q.; Hesler, J.L.; Weikle, R.M. On Estimating and Canceling Parasitic Capacitance in Submillimeter-Wave Planar Schottky Diodes. *IEEE Microw. Wirel. Compon. Lett.* **2009**, *19*, 807–809. [CrossRef]
20. Tang, A.Y.; Stake, J. Impact of Eddy Currents and Crowding Effects on High-Frequency Losses in Planar Schottky Diodes. *IEEE Trans. Electron Devices* **2011**, *58*, 3260–3269. [CrossRef]
21. Tang, A.Y.; Schlecht, E.; Lin, R.; Chattopadhyay, G.; Lee, C.; Gill, J.; Mehdi, I.; Stake, J. Electro-Thermal Model for Multi-Anode Schottky Diode Multipliers. *IEEE Trans. Terahertz Sci. Technol.* **2012**, *2*, 290–298. [CrossRef]
22. Siles, J.V.; Maestrini, A.; Alderman, B.; Davies, S.; Wang, H.; Treuttel, J.; Leclerc, E.; Narhi, T.; Goldstein, C. A Single-Waveguide In-Phase Power-Combined Frequency Doubler at 190 GHz. *IEEE Microw. Wirel. Compon. Lett.* **2011**, *21*, 332–334. [CrossRef]
23. Ding, J.; Maestrini, A.; Gatilova, L.; Cavanna, A.; Shi, S.; Wu, W. High Efficiency and Wideband 300 GHz Frequency Doubler Based on Six Schottky Diodes. *J. Infrared Millim. Terahertz Waves* **2017**, *38*, 1331–1341. [CrossRef]
24. Porterfield, D.W. High-Efficiency Terahertz Frequency Triplers. In Proceedings of the 2007 IEEE/MTT-S International Microwave Symposium, Honolulu, HI, USA, 3–8 June 2007.
25. Liu, H.; Powell, J.; Viegas, C.; Pérez-Moreno, C.G.; Alderman, B. A 40 to 160 GHz high power multiplier chain using planar Schottky diodes. In Proceedings of the 2015 8th UK, Europe, China Millimeter Waves and THz Technology Workshop (UCMMT), Cardiff, UK, 14–15 September 2015.
26. Chen, Z.; Wang, H.; Alderman, B.; Huggard, P.; Zhang, B.; Fan, Y. 190 GHz high power input frequency doubler based on Schottky diodes and AlN substrate. *IEICE Electron. Express* **2016**, *13*, 20160981. [CrossRef]
27. Yang, F. Discrete schottky diodes based terahertz frequency doubler for planetary science and remote sensing. *Microw. Opt. Technol. Lett.* **2017**, *59*, 966–970. [CrossRef]
28. Guo, C.; Shang, X.; Lancaster, M.J.; Xu, J.; Powell, J.; Wang, H.; Alderman, B.; Huggard, P.G. A 135–150-GHz Frequency Tripler With Waveguide Filter Matching. *IEEE Trans. Microw. Theory Tech.* **2018**, *66*, 4608–4616. [CrossRef]
29. Deng, J.; Yang, Y.; Zhu, Z.; Luo, X. A 140–220-GHz balanced doubler with 8.7–12.7% efficiency. *IEEE Microw. Wirel. Compon. Lett.* **2018**, *28*, 515–517. [CrossRef]
30. Dou, J.; Jiang, S.; Xu, J.; Wang, W. Design of D-band frequency doubler with compact power combiner. *Electron. Lett.* **2017**, *53*, 478–480. [CrossRef]
31. Zhang, B.; Fan, Y.; Zhang, S.X.; Yang, X.F.; Zhong, F.Q.; Chen, Z. 110 GHz high performanced varistor tripler. In Proceedings of the 2012 International Workshop on Microwave and Millimeter Wave Circuits and System Technology, Chengdu, China, 19–20 April 2012.

32. Hrobak, M.; Sterns, M.; Schramm, M.; Stein, W.; Schmidt, L.-P. Design and fabrication of broadband hybrid GaAs Schottky diode frequency multipliers. *IEEE Trans. Microw. Theory Techn.* **2013**, *61*, 4442–4460. [CrossRef]
33. *High Frequency Structure Simulator*; HFSS: Canonsburg, PL, USA, 2015.
34. *Advanced Design System*; Agilent Technologies: Santa Clara, CA, USA, 2013.
35. Erickson, N. Wideband High Efficiency Planar Diode Doublers. *Proc. Ninth Intl. Sypm. Space Thz Techn.* **1998**, 473–480.
36. Erickson, N. High efficiency submillimeter frequency multipliers. *IEEE Int. Dig. Microw. Symp.* **1990**, *3*, 1301–1304.

MDPI
St. Alban-Anlage 66
4052 Basel
Switzerland
Tel. +41 61 683 77 34
Fax +41 61 302 89 18
www.mdpi.com

Micromachines Editorial Office
E-mail: micromachines@mdpi.com
www.mdpi.com/journal/micromachines

www.ingramcontent.com/pod-product-compliance
Lightning Source LLC
LaVergne TN
LVHW070613170726
843515LV00004B/62

* 9 7 8 3 0 3 9 3 6 2 2 2 6 *